Management for Professionals

For further volumes:
http://www.springer.com/series/10101

Aristide van Aartsengel •
Selahattin Kurtoglu

A Guide to Continuous Improvement Transformation

Concepts, Processes, Implementation

Springer

Aristide van Aartsengel
Wijk Aan Zee
Netherlands

Selahattin Kurtoglu
Bochum
Germany

ISSN 2192-8096 ISSN 2192-810X (electronic)
ISBN 978-3-642-35903-3 ISBN 978-3-642-35904-0 (eBook)
DOI 10.1007/978-3-642-35904-0
Springer Heidelberg New York Dordrecht London

Library of Congress Control Number: 2013931849

Printed on acid-free paper

Springer is part of Springer Science+Business Media (www.springer.com)

Acknowledgments

Throughout the two books entitled: "A Guide to Continuous Improvement Transformation: Concepts, Processes, Implementation" and "Handbook on Continuous Improvement Transformation: The Lean Six Sigma Framework and Systematic Methodology for Implementation," we have referenced many illustrious practitioners to whom we are obviously indebted. The text of these volumes books has also been hugely improved through the friendly criticism and advice given by various extremely generous individuals.

We would like to express our gratitude to all those who taught us, worked with us over the years, and those who have helped us with this work or inspired new ideas. They are, literally, too numerous to mention. Many of the ideas and examples come from practice. We are therefore especially indebted to the many colleagues, managers and CEOs who have allowed us to share their work on continuous improvement and project management.

We also wish to acknowledge dozens of people from our client organizations, practicing Kaizen in manufacturing plants, Business Process Management, Project Management, Lean Six Sigma, Lean Manufacturing, Total Quality Management (TQM), Total Quality Control (TQC) and Total Productive Maintenance (TPM) to whom we owe special thanks and who have shown the applicability of the ideas and methods described in this handbook.

We would also like to acknowledge all the client organizations over the years that have trusted our advice and provided us with the greatest laboratory there is – their organizations. Their willingness to test new hypotheses contributed greatly to the material. We extend a deep bow to IQPM Consulting for giving us such an interesting subject about which to learn.

Finally, our families deserve loving mention, and sincere thanks, for putting up with the hours of time spent hunched over our computers writing and revising the content of this book.

Aristide van Aartsengel, Ph.D., Six Sigma Master Black Belt
Selahattin Kurtoglu, Ph.D., Project Management Professional

Contents

List of Figures

List of Tables

Introduction 1

An enterprise business exists for one purpose: to create value and generate revenues in order to continue to exist in today's ever changing business landscape. These revenues must cover the expenses of the enterprise business, provide working capital for future operations, and provide the profits expected by the shareholders.

Because of this economic business imperative, all managers have to focus on the bottom line of the cash flow statement. Hence, from an enterprise business executive's point of view the enterprise business world consists of three things: income, expenses, and profits. Furthermore, the objective for the enterprise executive is clear: increase income, lower expenses, and maximize profit in the short term.

Therefore, the most challenging question confronting enterprise business leaders and managers in this new millennium is "How can an enterprise business executive and its business achieve this objective in the short term in order to continue to exist in the long term?" In other words, the question confronting enterprise business leaders and managers is not "How do we succeed?" It is: "How do we stay successful?"

1.1 The Issue

Countless books, papers, training seminars proceedings and missives in the past 25 or more years have been written passionately about successful management practices and techniques developed by Toyota and GE family of companies and presented to management to improve organizational performance. But strangely enough, despite the vast amount of knowledge presented to management on these much heralded set of management practices, no organization outside the leaders enterprise businesses that are Toyota's and GE's family of companies has ever come close to matching Toyota's or GE's top performances. The prevalent feeling is that something these leaders enterprise businesses do is still not understood and put into practice by other enterprise businesses.

Toyota and GE management practices and techniques presented to management to improve organizational performance all sound good and programs to implement each of them produce results if enough money is invested and the enterprise business

A. Van Aartsengel and S. Kurtoglu, *A Guide to Continuous Improvement Transformation*, Management for Professionals, DOI 10.1007/978-3-642-35904-0_1,

executive management gives full and total support to it. The problem occurs because in many enterprise businesses today, less than 20 % of the total enterprise business' budget can be set aside for the implementation of performance improvement practices and techniques. Approximately 80 % of the budget needs to go to pay salaries, buy materials, pay taxes, pay shipping costs, etc. Even if there was enough cash flow to invest in all of these performance improvement practices and techniques, there is no way that management could make numerous performance improvement practices and techniques their top priority, as you can only have one top priority. The result is that the enterprise business top management listens to everyone's pitch in search for the secret of success in performance improvement and then spreads the cash flow around, giving the biggest share of it and the executive's time to the enterprise business that makes the best presentation. This means that none of the presented improvement practices and techniques gets enough cash flow to undergo a successful implementation.

Most initiatives and programs to improve organizational performance fail to deliver. Whether the aim is to transform an enterprise business to become effective and efficient, more productive, customer intimate, or growth oriented, the majority of these organizational initiatives and programs fail to accomplish their goals. Most of these efforts do not lower expenses, improve productivity, boost customer satisfaction, or raise revenue to the levels that the enterprise business' executive board expected or promised to its stakeholders. Some initiatives and programs, in fact, fail entirely. All in all, the track record of business improvement initiatives and programs is a sorry one.

Why do enterprise businesses fail to achieve the bottom line of increasing income, lowering expenses, and maximizing profit in the short term while maintaining a consistency of purpose (i.e. sustaining results to foster the enterprise business mission and sustaining the business in the long term) despite the vast amount of knowledge on Toyota and GE management practices and techniques?

Because the competitive advantage of an enterprise business lies not so much in replicating Toyota and GE management practices and techniques themselves – whether it be 'lean' techniques, six sigma, today's profitable product, or any other – but in the ability of the enterprise business to understand conditions and adapt to create fitting, smart solutions. Toyota and GE management practices and techniques are solutions developed for the specific problems that Toyota and GE family of companies were facing.

The much heralded superior results of GE and Toyota family of companies spring more from patterns of 'continuous improvement' via experimentation than from the management practices and techniques that have been reported in the literature. Many of those tools and practices are, in fact, countermeasures developed out of 'continuous improvement' invisible patterns of thinking and acting. Focusing on implementing these solutions into your business will not make your enterprise adaptive and successful. These solutions are built upon invisible patterns of thinking and acting, and consistent strategic improvement and adaptation.

Achieving and sustaining success in today's hyper-competitive marketplace, with severe economic turmoil, is an ultimate challenge for any company and business leader. This is an era of unprecedented change, complexity, volatility, and risk when everything seems to be moving at very fast speed. There is very little room for error. The business imperative today is not just to perform excellently, but to perform excellently consistently. The good thing about the many books, papers, training seminars and missives on improvement methodologies that are presented to management is that all of these methodologies have a lot of things in common. Activities like team building and project management are common in all the methodologies, so if the enterprise business focuses upon the common activity, the impact will be the greatest on the enterprise business' performance. Consequently, what enterprise business leaders want to do is to optimize the enterprise business' short- and long-range performance as viewed by all of the stakeholders.

1.2 The Solution

While the objectives of increased revenues, reduced expenses, and increased profits are legitimate in the short term, the real question is always "how these goals will be met in the short term and sustained in the long term?" The answer is "Continuous Improvement" transformation, a concept that is popular today, as alternative is to "Continuous Beatings" with stuttering failure of the traditional business approaches where you constantly strive for that magical combination of top talent, superior management, critical resources and canny strategy. But those elements – separately or in combination – will never achieve their ultimate value without ongoing and step wise improvement initiative in place.

The only reason why you, as an enterprise business executive, should ever consider starting a "Continuous Improvement" transformation initiative is to generate more profits and make your enterprise business more competitive. Do not start a "Continuous Improvement" transformation initiative to improve customer satisfaction or employee morale. It will do that, but the real reason you need a "Continuous Improvement" transformation initiative is to increase profits in the short-term and sustain these gains in the long-term; hence, you should look at the "Continuous Improvement" thrust as a business investment that is either going to add to or detract from long-term, net favorable balance.

1.3 So You Want to Start a "Continuous Improvement" Initiative

"Continuous Improvement" transformation is not optional; it is not part of the game. It is the game today; a condition of survival. For an enterprise business to survive in today's competitive international environment there must be improvement efforts in both the continuous improvement philosophy and break-through improvement methodology. Every enterprise business must have systematic methods for making smart decisions, attacking problems, improving its products (i.e. tangible products) and services (i.e. intangible products), repelling competitors, and keeping customers delighted. Anything less than a systematic, disciplined approach is leaving the enterprise business future in the hands of chance.

"Continuous Improvement" transformation initiatives and programs have tallied billions in savings, dramatic increases in speed, strong new customer relationships – in short, remarkable results and rave reviews. What is "Continuous Improvement" transformation? What are its key characteristics and constituents? And how do we take the enterprise business to the next level and realize "Continuous Improvement" transformation?

1.4 The Purpose of This Book and Our Next Book

Most enterprise businesses do not progress beyond an average level of effectiveness regarding their ability to improve continuously, relative to Toyota and GE family of companies or to other enterprise businesses in their respective industry. Most continuous improvement initiatives are stuck at an immature level of development, in that they merely help the enterprise business to survive – they do not transform the enterprise business' competitive position. The goals of this book are:

1. To provide a clear, well structured and interesting account of the concept of "Continuous Improvement" transformation as it applies to a variety of successful businesses and organizations;
2. To enable enterprise business leaders – from CEOs to supervisors – understand what "Continuous Improvement" transformation is, both a simple and a complex question, why it is probably the best answer to improved business performance in years, and how to create an optimal environment to maximize your chances for successful implementation work in the context of your enterprise business.

The text provides a logical path through the activities of a range of overarching determining factors of strategic management that matter the most. It aims to be:

1. *Strategic in its perspective*, it is unambiguous in treating the following overarching determining factors of strategic management as being central to competitiveness:
   - Leadership
   - Culture and Values
   - Strategic Planning and Management
   - Performance Measurement
   - Performance Management
   - Alignment and Commitment
   - Process Improvement and Management
   - Sustainability
2. *Conceptual* in the way it explains the reasons why "Continuous Improvement" is a condition of survival.
3. *Comprehensive* in its coverage of the significant ideas and issues which are relevant to achieve a state of "Continuous Improvement".
4. *Practical* in that it provides guidance needed to create an optimal environment for realizing "Continuous Improvement" in practice.

The eight overarching determining factors of strategic management selected are drivers necessary to achieve a state of "Continuous Improvement" as developed in this book. They were designed by The Balanced Scorecard Institute in its white paper on Strategic Management Maturity Model (SMMM) (BSI, 2010) to:

1. Help enterprise business leaders – from CEOs to supervisors – perform a quick assessment of where their enterprise business stands in terms of strategic management;
2. Monitor progress in improving maturity of their enterprise business; and
3. Allow benchmarking across organizations, or departments within one organization, in order to identify best practices.

These determining factors are either missing or weakly applied in most enterprise business performance-improvement recipes. They are not new or probably even surprising, but they are critical to providing a solid foundation for "Continuous Improvement" transformation. Entire books have been devoted to each one of these overarching determining factors of strategic management. Based on our experience with our client organizations and on research conducted, we have concluded that many enterprise businesses do not implement these overarching determining factors well.

These eight overarching determining factors are not hard-wired into most improvement initiatives. If an enterprise business wishes to accomplish more with its investments in "Continuous Improvement," it should include them in its recipe. They can differentiate the enterprise business from the competition and help to accomplish more meaningful change. They are further carried out in details throughout this book, informing enterprise business leaders – Project Managers, Green Belts, Black Belts, Master Black Belts, managers at all levels, and process improvement professionals – on the issues that matter most to success.

1.4.1 What Makes This Book Different

The distinctive feature of this book that makes it different from the remaining literature is that it addresses thoroughly the basic question to ask of management for a successful implementation of any improvement initiative: "Are we doing the right things?"

This book, the one that you are reading now, is not about a framework and systematic methodology for studying the constituent elements or processes and systems associated with each of the eight overarching determining factors. Such a framework and systematic methodology focus on answering the question "Are we doing things right?" This book is about the strategic management and the social environment concerns, the chapters of which answer the question, "Are we doing the right things?" It is about the management thinking and practices carried out throughout the eight overarching determining factors, and which must be put to practice on a daily basis.

A hammer and saw do not make someone a carpenter. Nor does mastery of technical skills alone ensure success. In any enterprise business, it is the intended strategy, driven from the vision of the leadership that defines what the right things are. Thus, a systematic improvement methodology like 'Lean Six-Sigma,' or any other quality improvement methodology, alone cannot guarantee that an enterprise business will be successful, that it will achieve its intended mission, or that it will progressively realize its full potential.

These two aspects of management – strategic and operational – complement each other, so both must be assessed to determine the enterprise business' capabilities to improve continuously. Consequently, we have written a separate book, the topic of which addresses operational deployment and management of "process improvement" projects by integrating the project management precepts covered in what has emerged as the world standard of project management knowledge – A Guide to the Project Management Body of Knowledge (Project Management Institute) – with the Plan-Do-Study-Act (PDSA) model for improvement and the Lean Six-Sigma concepts.

Once you have read the present book, you will probably be surprised at how simple the basic concepts on "Continuous Improvement" are, but performing these concepts right is not necessarily easy. That is why this book focuses on the most important and highest-leverage strategies, tactics, and action proposals you can use for releasing the power of your enterprise business "Continuous Improvement" capabilities to achieve transformational results.

Our hope is that this book provides a comprehensive resource not only for those in corporate, but also for individuals in the public sector; continuous improvement practices for governmental organizations are inarguably equal in importance to such practices in private industry.

Enjoy this book. Take it seriously. Put it to work, and over time, what was stressful will become merely interesting, what you avoided will become attractive, and what seemed futile will become a source of possibility.

Defining 'Continuous Improvement' 2

This chapter is concerned with the core of the art of "Continuous Improvement" transformation and delves into the key characteristics and constituents necessary to take the enterprise business to the next level to continue to exist in the long term. Subsequent chapters provide guidance to enterprises management and to professionals engaged in the "Continuous Improvement" initiative implementation and enable them to structure and manage its implementation successfully.

2.1 Setting the Stage

As an enterprise business executive, the only reason why you should ever consider starting a "Continuous Improvement" transformation initiative is to generate more profits in the short-term, sustain these gains in the long-term and make your enterprise business more competitive. If a "Continuous Improvement" transformation initiative is launched to improve customer satisfaction or employee morale, it will do that. But you should look at the "Continuous Improvement" thrust as a business investment that is either going to add to or detract from long-term, net favorable balance of the profit formula. The profit formula is the blueprint that defines how your enterprise business creates value for itself while providing value to the customer. It consists of the following:

1. *A Revenue Model*, which is equals the price required to deliver the customer value proposition (CVP) times the respective volume of products and services. The volume can be though of in terms of market size, purchase frequency, ancillary sales, etc.
2. *A Cost Structure:* direct costs, indirect costs, economies of scale. Cost structure will be predominantly driven by the cost of the critical resources required by the business model. The critical resources are assets such as the people, technology, products, facilities, equipment, channels, and brand required to deliver the value proposition to the targeted customer. The focus here is on the key elements that create value for the customer and the company, and the way those elements

A. Van Aartsengel and S. Kurtoglu, *A Guide to Continuous Improvement Transformation*, Management for Professionals, DOI 10.1007/978-3-642-35904-0_2,

interact. Every enterprise business also has generic resources that do not create competitive differentiation.

3. *A Margin Model:* given the expected volume of products and services and cost structure, the contribution needed from each transaction to achieve desired profits.
4. *A Resource Velocity:* how fast you need to turn over inventory, fixed assets, and other assets – and, overall, how well you need to utilize resources – to support your expected volume of products and services and achieve your anticipated profits.

Traditionally, enterprises executives and managers commonly seek to meet the goal of "increase income, lower expenses, maximize profit" in the short term by setting sales goals, establishing quotas and targets, launching advertising campaigns, creating new products or packaging old products in new ways, or raising prices to whatever the market will bear. At the same time, in order to lower expenses the executives and managers decrease the workforce, cut back on inventory levels, provide lower levels of service, and slash all indirect expenses.

As a result, the enterprise as a whole (businesses and customers), is trapped with decreasing levels of service and with a growing sense that everyone is increasingly harried as employees are left to do more work with fewer resources. Within the enterprise, every department, and often every individual within a department, is in competition for scarce internal resources, and cooperation is rare. As individual within the enterprise scrambles to take care of his or her own turf, the business suffers, and so new waves of cost reductions are put in place, continuing the downward cycle.

When the goal of "increase income, lower expenses, and maximize profit" is not met at the corporate level, the corporate business units get new goals and performance targets. When the business units' goals and performance targets are not met, the departments get new goals and performance targets. And when the department goals and performance targets are not met, the manager is reprimanded and individuals are given "stretch goals." In the long term, this leads to a cover-your-anatomy mentality in which the need to survive the internal competition becomes dominant, taking precedence over the needs of the business and the needs of the customer. The enterprise often becomes little more than a collection of assets to be reorganized, stripped, sold, and resold, with decision-making driven by individual advocacy and self-interest. Sound familiar?

Why do most enterprise business executives and managers so often choose to make decisions that systematically decrease the long-term value of their businesses? One reason may be that they appear to be captives to the "zero sum"[1]

[1] In economic theory, "zero sum" thinking is a representation of a situation in which a participant's gain or loss is exactly balanced by the losses or gains of the other participant(s). If the total gains of the participants are added up, and the total losses are subtracted, they will sum to zero. Cutting a cake is a zero-sum situation, because taking a larger piece reduces the amount of cake available for others. In contrast, non-zero-sum describes a situation in which the interacting parties' aggregate gains and losses is either less than or more than zero.

thinking, thus most enterprise business executives and managers define their enterprise businesses' interests too narrowly.

If these enterprise business executives and managers define the enterprise business' self-interest (and consequently its goals) too narrowly – for example, to maximize this year's or this quarter's reported earnings – they will view that interest as being at odds with the interests of their customers and employees. From that perspective, in the short term every unit of money spent on employee training is viewed as a unit of money of lost profit. Every additional unit of money squeezed out of a customer, even if it comes at the cost of poor service or price gouging, improves this quarter's results.

This narrow view is powerfully reinforced by financial accounting systems that were well adapted to the industrial economy, but are inadequate in today's economy. The accounting and finance conventions of the industrial age are good at valuing tangible assets, but they largely ignore the value of harder-to-quantify or intangibles assets like employee satisfaction, learning, R&D effectiveness, customer loyalty, etc. In today's economy, those intangible assets are far more important than the assets that traditional accounting systems were designed to measure.

Enterprise businesses which operate with the "zero sum" thinking of self-interest may stumble into a downward spiral of poor decision-making, which is difficult to reverse. For example, as illustrated in Fig. 2.1, as reduced employee training and compensation lead to low employee morale and poor performance, and as underfunded R&D allows a product line to age, customers can become dissatisfied and begin to defect.

In situations where customers are "locked-in" owing to large investments in proprietary equipment or some other temporary monopoly effect, they may not defect immediately. Instead, they will become increasingly alienated and defect as soon as a product or technology shift, regulatory change, or competitive offering allows it. When customers finally do defect, profits shrink, tempting management to cut back even further on training, compensation, and R&D, thus accelerating the spiral of customer and employee dissatisfaction and defection.

Here are just a three of the areas where zero-sum thinking rears its consequences in the enterprise business arena:

1. *Squeezing suppliers* – In the quest for cost-cutting, enterprise business executives and managers have focused on squeezing the prices of their suppliers as much as possible. The result has been deteriorating trust and relationships with key business partners. Too many enterprise business executives and managers under-estimate the opportunity of working together to make both parties stronger and deliver even more value to the marketplace.
2. *Growing focus on intellectual property protection* – There are certainly valid concerns here, but too often enterprise business executives and managers seek to protect their existing stocks of knowledge at the expense of the opportunity to participate in broader relationships that could significantly refresh these stocks.
3. *Marginalizations of innovation* – With some obvious exceptions, large enterprises have generally become consumed with the quest for cost-cutting – again, for understandable reasons. In the process, though, the opportunity to create

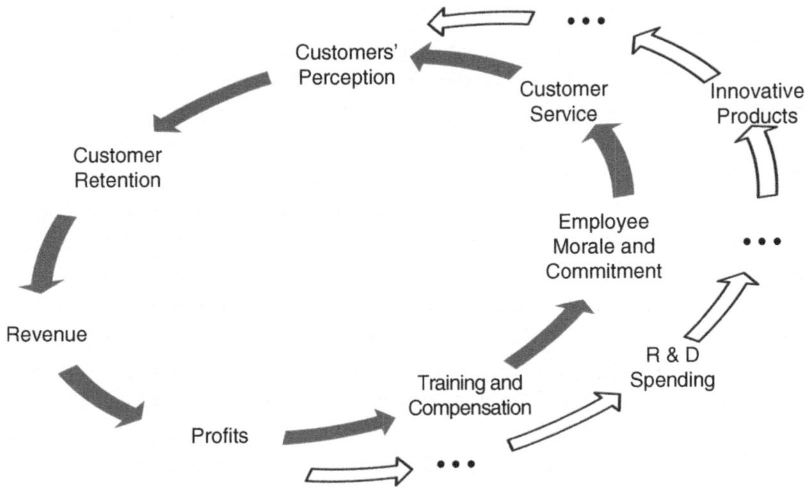

Fig. 2.1 Path for generating customer defection

new forms of value through innovation has been shunted aside. Innovation has been placed in compartment into R&D departments that have been squeezed for cost-savings along with everyone else.

Rather than assigning innovation to the ghetto of R&D, why not liberate innovation and view it as an activity that everyone in the enterprise should be pursuing every day? Of course, that means breaking the mindset that innovation is about product development. After all, innovation is ultimately about finding ways to deliver new value to the marketplace from existing enterprise resources, whether this value is in the form of products, new work practices, improved business processes, new management techniques or new business models.

This cycle of destruction is seen in many enterprise businesses today. In fact, most initiatives to improve enterprise business performance fail to deliver. Whether the aim is to increase income, lower expenses, or maximize profit, the majority of initiatives to improve enterprise performance fail to accomplish their goals. Most of these efforts do not reduce costs, improve productivity, increase customer satisfaction, or raise revenues to the levels that executives expect or have promised to their stakeholders. Some initiatives, in fact, fail outright because of a blindness ignorance and a lack of understanding of the constituents required to improve performance in an ongoing basis and consistently. Under the zero sum thinking, the track record of improvement initiatives within most enterprises businesses is a disastrous one.

Alternatively, if enterprise business executives and managers define the enterprise business' self-interest (and consequently its goals) broadly enough to include the interests of customers and employees, an equally powerful spiral of value creation can occur. Highly motivated, well-trained, properly rewarded employees deliver outstanding service, while effective R&D investments lead to products that enjoy a significant value-adding advantage and generate higher margins. Satisfied, loyal customers (and new customers responding to word-of-mouth referrals) drive

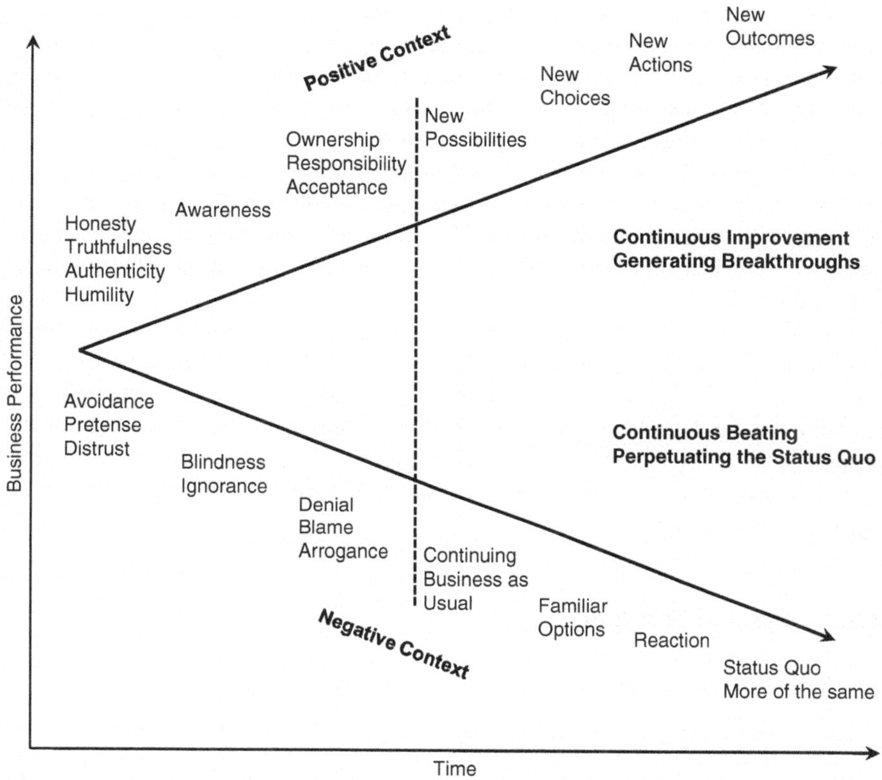

Fig. 2.2 Two paths: generating breakthroughs through continuous improvement versus perpetuating the status quo through continuous beating

revenue growth and profitability for shareholders. Clearly, the undesirable reinforcing processes described in Fig. 2.1 can work in reverse.

The business imperative in these times of severe economic turmoil is not just to perform excellently, but to perform excellently consistently. While the objectives of increased revenues, reduced expenses, and increased profits are legitimate in the short term, the real question is always "how these goals will be met in the short term and sustained in the long term?" The answer is "Continuous Improvement" as alternative is to "Continuous Beatings" with stuttering failure of the traditional approach described above and illustrated in Fig. 2.2.

2.2 What Is "Continuous Improvement"?

How do we define "Continuous Improvement" and what are its key characteristics and constituents? Ideally we need a clear definition of "Continuous Improvement" in order to clearly understand what it is, how it differs from other constructs, what it is related to, and how it should be measured.

To address these questions, we adopt the systems theory framework throughout the remaining of this book for viewing the enterprise as a whole. We also use the concept of maturity which is popular today, with new models emerging describing many aspects of individual maturity, professional maturity, team maturity, process/program/project maturity, and enterprise maturity.

An enterprise, by its most basic definition, refers to an assembly of people working together to achieve common objectives through a division of labor. An enterprise provides a means of using individual strengths within a group to achieve more than can be accomplished by the aggregate efforts of group members working individually.

Enterprise businesses are formed to create value through the provision of goods or services to consumers in such a manner that they can realize a profit at the conclusion of the transaction. Over the years, business analysts, economists, and academic researchers have pondered several models that attempt to explain the dynamics of enterprise businesses, including the ways in which they make decisions, distribute power and control, resolve conflict, and promote or resist organizational change.

2.2.1 System Thinking

We shall think of a system as:

> A deterministic entity comprising an interacting collection of discrete elements.

From a practical standpoint, it is necessary to specify further what aspects of system performance are of immediate concern. A system performs certain functions and the selection of particular performance aspects will dictate what kind of improvements are to be conducted. For example: we are interested in whether the system accomplishes some task successfully; are we interested in whether the system fails in some unpredictable way; or are we interested in whether the system will prove more costly than originally anticipated?

The "deterministic" nature of the entity considered implies that the 'system' in question be identifiable. It is entirely ineffective to attempt to improve performance of something that cannot be clearly identified. Furthermore, a system must have some purpose – it must do something. Enterprise businesses, organizations, school systems all have definite purpose and do not exist simply as figments of the imagination.

The "discrete element" of the entity considered must also, of course, be identifiable; for instance, the individual business unit of an enterprise business. Note that the discrete element themselves may be regarded as systems. Thus, a business unit of an enterprise business consists of departments, line groups, and so forth; each of these, in turn, may be further broken down into subsystems, etc.

Note also from the definition that a system is made up of parts or subsystems that interact. This interaction, which may be very complex indeed, generally insures that a system is not simply equal to the sum of its parts. Furthermore, if the performance of any part changes – for example any type of failure – the system itself also changes. This is an important point because, should determining factors changes be made as a result of a system improvement, the new system so resulting will have to be subjected to an improvement of its own.

Of the various organizational models that have been put forward in this realm, the "system thinking" has emerged as the most widely known. System thinking focused on overall system properties and characteristics which appeared to apply generally to all living systems from a simple biological cell to a complex social organization such as a business. These critical system properties appeared capable of providing an overall explanation of system behavior and this was used in the analysis of industries' chronological development as well as product life cycles.

Koehler candle's life cycle analogy provides a good and very simple explanation of the performance of system (Koehler, 1938). At first when a light is put to its wick the candle may spit and sputter and possibly go out several times before the wax achieves the temperature for ignition and is successfully lighted. This birth, introduction or infancy stage is characterized in many systems by volatility and high rates of infant mortality whether we are considering lighted candles, human babies, electronic components, business start-ups or new products.

If the candle successfully lights then the flame quickly burns up to its full size. This adolescent or growth stage is again typical of many systems in the speed and continuity of its growth up to a certain ceiling level characteristic of its mature phase.

As the candle reaches this ceiling it exhibits a generally applicable characteristic of volatility before settling down to a "mature"-phase steady state. The candle's volatility is manifested in a short period of flickering; adolescent human beings exhibit extraordinary volatility as any parent will vouch; the volatility of enterprise businesses and products as they move from growth to maturity is also remarkable, for example, as growth predictions have to be permanently downgraded, forcing management's attention on to a different set of problems.

In its "mature"-phase steady state, the candle exhibits the general systems characteristic of maximum strength, effectiveness and efficiency. This is the phase when the candle burns the wax fastest and gives off the greatest light. It will maintain this maximum energy conversion steady state as determined by its inputs of wick, wax and oxygen and its system characteristics of size and composition of wick and diameter and length of candle. Human beings exhibit similar characteristics, in their "maturity" being at their strongest and physically most efficient stage. Enterprise businesses also appear to be at their most efficient and intrinsically most profitable, cash generating stage.

The steady state will only end when one of the determining factors changes. For example, the wax is used up to the extent that there is no longer a full quantity available for burning. At this stage, the system goes into a decline, but again the change from the "mature" phase to decline is marked by further volatility as the candle flickers and putters and frequently goes out prematurely, i.e. before it has used up all its wax.

One of the intriguing characteristics of this general model is the apparent breadth of its applicability. All manner of systems appear to share these general characteristics and be subject to parallel pressures and influences at the different stages.

A start-up enterprise business, for example, is likely to be dominated initially by the need to survive. If it survives this first phase, it will be able to turn its attention to growth and the development of competitive strength. As it progresses through adolescence it is quick, flexible, opportunistic, and focused on satisfying its shareholders and its customers' needs. It carries no spare weight, no passengers. It is lean and fit, quick on its feet and builds its strength through constant striving and exercise.

This phase sees the enterprise business change from being the creature of the founding entrepreneur with a simple structure, to employing an increasing number of professional specialists concerned with either the technological development or the development of its various management functions.

In a growing market, the adolescent enterprise business has to run fast in order simply to maintain its market share. If it fails to do this, it will probably not survive the first shake out when market growth starts to weaken. In a static market, there is not the same necessity to grow. Many enterprise businesses remain small, providing relatively stable employment for small numbers of people. Nevertheless, other businesses are more ambitious and grow rapidly in order to achieve the critical mass at which the new specialists can be profitably supported. Growth in static markets can only be achieved by increasing market share or by moving into new markets, both of which may be problematic in highly competitive situations.

To achieve "maturity" is the goal of all systems. In the case of an enterprise business, "maturity" is the phase when value creation and wealth creation is maximized, when the most surplus cash is generated, when the enterprise business achieves its position of greatest power and influence, and when the enterprise business should be able to focus, with the least inhibition and interruption, on the achievement of its long-term intended objectives.

Maturity connotes experience, wisdom, and effectiveness. An enterprise that is maturing is getting better – much better. Maturity is the result of successful infancy and adolescence. The success is usually based on doing the right things right and the enterprise business progressively becoming more expert. It learns successful ways of doing things. It finds out what its shareholders and its customers like and gets good at delivering those things. It develops its technological expertise. It uses recipes which work and it becomes efficient. And it becomes effective. All this happens as a result of deliberate management initiatives – it is by no means automatic.

On the human development literature, there are concepts that are very clearly associated with maturity. They include: the development of wisdom; the eagerness to confront reality; the need to learn from the past; the will and wish to act independently; the need to know when to conform; the ability to adapt to ongoing change; the need to remain open to new ideas; the willingness to question one's own belief system; the aptitude for not being threatened by questions from others; and many more.

For a successful enterprise business, however, it is difficult to make a fundamental change in what has established its leading position. This is especially the case with any technology when a successful mature enterprise business is likely to have major investments sunk in the old technology. Getting into something new may mean writing off huge capital assets which will weaken the balance sheet and in the short term wreak havoc with profitability. Also, there are psychological investments in the old technology.

One of the fruits of "maturity" is the ability to pay top salaries and thus attract top highly qualified people. Many of these highly qualified professionals may have built their entire careers on the old technology and their very natural response to such a change is likely to be defensive and reactive. Nor is it at all certain that leadership in the new technology will necessarily follow being a leader in the old; giving up a leading position should certainly not be done lightly.

A successful mature enterprise business, as the system model suggests, is likely to generate substantial surplus funds which are not required in order to maintain the status quo. How these funds get invested depends very much on the circumstances of the individual enterprise business.

Like all organisms, an enterprise business exists primarily for its own survival and improvement: to fulfill its potential and become mature or as great as it can be. In this perspective, we can think of maturity as the progressive realization of the enterprise full potential. It is a process of a continuously innovative course of actions of improvement, introducing the new, eliminating waste, reducing costs, improving quality, and so giving the customer a better deal than competitors do. This is the mindset which drives an enterprise business management across all its responsibilities in operations, marketing, finance and technical management, focusing on ever better use of resources so as to ensure the enterprise business' survival and its ability to win in its chosen markets.

Accordingly, the progressive realization of the full potential of the enterprise business provides the possibility to continue to develop beyond anything that is currently known. There is no final end of the road to this progressive realization of the enterprise business full potential. However, there is a state of "becoming" – becoming more relevant; becoming more functional; becoming more powerful.

Thus, we define a state of "Continuous Improvement" as any state of "being" beyond the state of "becoming." It is the highest stage of maturity maturity that an enterprise business as a whole can attain. Attaining this highest stage of maturity does not happen overnight; it takes time!

Most of our client organizations think of "Continuous Improvement" transformation as something that happens periodically, like a project, a workshop, or campaign: they make a special effort to improve or change when the need becomes urgent. But this is not how "Continuous Improvement" transformation, adaptation, and sustained competitive advantage actually come into being. Relying on periodic improvement or change efforts should be seen for what it is: only a sporadic add-on to a system that by its nature tends to stand still.

As illustrated here above, the system approach offers a way of understanding enterprise businesses. Enterprise businesses and particular organizational situations can be analyzed in terms of the various interacting systems they comprise. These may be social, organizational, or technological. System thinking provides the basis for a structured and consistent way of thinking and managing an enterprise business to improve performance, and yet, allows for creativity and adaptation. Creativity and adaptation must always be built into the system and ad-hoc decisions can be taken when the need arises.

The system approach to transformation builds on an understanding of the interactions and interdependencies within and across the constituent subsystems. It highlights the ability to analyze subsystems interconnections, identify system improvement opportunities, and create strategies to translate those opportunities into value.

Adopting a system thinking approach makes the impacts of various components of an enterprise business as a whole more visible and thus more manageable. For example, taking a system view may reveal that the root cause of why the enterprise business cannot meet its cost, quality, and delivery objectives is not something that lies exclusively in any one of the functional areas of marketing, material acquisition, manufacturing, or logistics, etc. but rather in the integration across all its constituent functional areas.

System thinking should happen at all levels of an enterprise business: at the strategic and operational level as well as the interaction between them. The Deming circle of Plan – Do – Study – Act, is an example of this. System thinking would suggest that in the progressive realization of the enterprise business full potential, management creates a 'Plan' of what it would like to 'Do'. This goes into the Execution mode that may either resemble a process, a project or smaller set of activities. The outcome of the 'Do' step needs to be 'Studied' and/or 'Monitored' over time and at the completion of the activity. As a result of the outcomes, there will be triggered a need to take action (i.e. Act). The following are four situations that make the achievement of systems thinking and acting difficult or impossible to achieve.

1. *You cannot achieve your target, unless you manage it* – Targets and goals are rarely met without the involvement of "management" and management action. If the targets are met without management involvement, then they simply were not ambitious enough. Management provides guidance and ensures that the various pieces of the puzzle fit together. Management requires clear definitions of roles and responsibilities, including ownership.
2. *You cannot manage what you do not measure* – Management requires measurement. While the popular "management by walking around" is an important tool to gain a sense of what is happening "on the workshop floor", it can never be the only tool, nor replace true measurement of process and people performance.
3. *You cannot improve without management* – There are still many enterprise business that have a low level of business process management maturity and yet still attempt to start business process improvement activities without firstly establishing the management required for these processes. Even if the enterprise business does achieve some process improvement, the gains will rapidly

disappear unless the business processes are managed for sustainability. In our experience, many Lean Six Sigma initiatives fall into this category.

4. *No alignment without governance* – Process governance is critical to the systems thinking figure. Process governance must ensure that the target, execution, management and improvement activities are aligned. This is crucial as the various roles for these aspects are distributed among different people within the enterprise business. A pragmatic approach to process governance within an enterprise business will increase the commitment and adherence of all concerned.

2.2.2 Characteristics of Enterprise Business Maturity Stages

An enterprise business maturity determines the enterprise business ability to continuously improve, and is a key determinant of its future performance. Although some maturation occurs naturally through normal learning and experience, it is generally accepted that systematic development interventions will enable the attainment of higher stages of maturation more quickly.

Research also indicates that only 20 % (one in five) of organizational efforts, for progressive improvement toward the "Continuous Improvement" stage of maturity, achieve long-term sustainability. We believe that this is because there is rarely an understanding of the determining factors for real maturation. One of the main reasons we wrote this book is to help enterprise businesses and organizations accelerate the movement from their actual maturity stage to higher stages of maturity.

As illustrated in Fig. 2.3, progressive realization of the full potential of the enterprise business generally undergoes five stages. These are:

1. Ad Hoc and Static stage – Disaster
2. Reactive stage – Learning Tools
3. Structure and Proactive stage – Tools Master
4. Managed and Focused stage – Transformation, and
5. Continuous Improvement stage – Winning Team Culture.

2.2.2.1 Stage 1: Ad Hoc and Static: Disaster

The lowest stage of maturity, "Ad hoc & static," characterizes enterprise businesses that do not have any strategic planning or management in a formal sense, tending to plan in an ad hoc and uncontrolled manner, normally by senior management behind closed doors, and never addressing long-term strategy. Such enterprise businesses were organized for the industrial era, utilizing command and control orientations that are inadequate for today's environment. At this stage, there is relatively little performance measurement above and beyond what is legally required. Any performance measurement that does exist can best be described at sporadic and unplanned.

Thus, a "Stage 1" maturity enterprise business is a disaster. There are so many problems in an enterprise business at this level that no one has the time or interest in attempting a performance improvement initiative. One can argue that they should, but typically this would not be perceived as the most important action currently

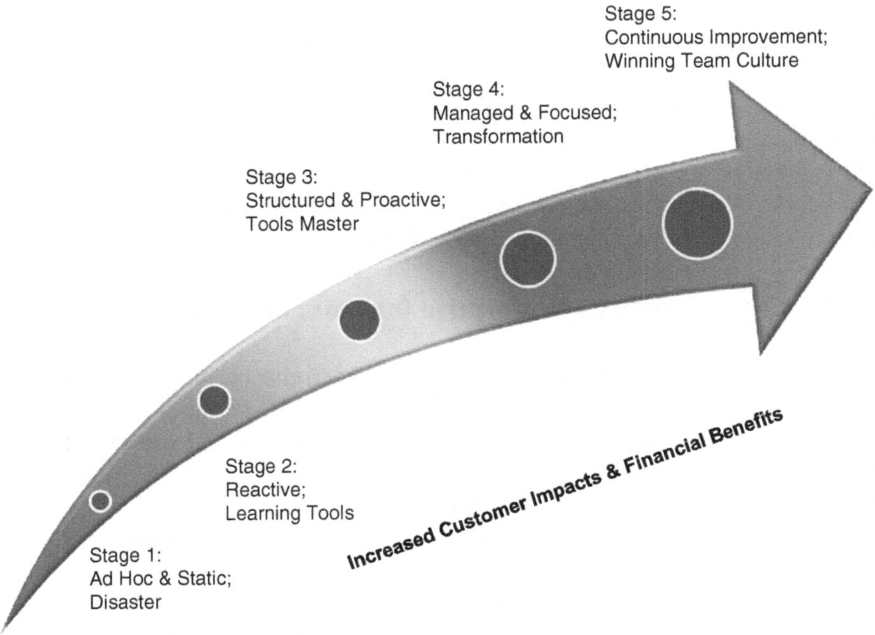

Fig. 2.3 Enterprise business progressive maturity stages

needed. "Fires need to be extinguished first," would be the typical reaction within such enterprise businesses. This category includes enterprise businesses that are significantly underperforming their industry.

2.2.2.2 Stage 2: Reactive: Learning Tools

The second stage of maturity, "Reactive," characterizes enterprise businesses that have developed some elements of effective planning with strategic performance management being applied, only in an inconsistent fashion and often with poor results. Planning discipline is unlikely to be rigorous, and only happens in reaction to events or to temporarily please a specific individual within the enterprise. Enterprise businesses at this stage of maturity might measure performance and performance measures might be used by enterprise managers merely to get rewards and to punish underperformers.

Industry performers at this level are usually just learning an improvement methodology or they are going back to improvement basics. The typical sequence of events within enterprise businesses at this stage of maturity is as follows:

1. Someone with influence in a "Stage 2" enterprise business decides an improvement need exists, or an executive becomes captivated with a new methodology that he or she has discovered.
2. Then, a Champion of Improvement is appointed to lead the enterprise business' initial efforts at learning and using the relevant tools.

3. Training events teach employees about the improvement tools (e.g., 5S, Value Stream Mapping, Statistical Process Analysis, Lean, Six Sigma, supply-chain management, etc.).
4. Then the enterprise business embarks on a journey to prove the tools work in their environment.

Also, the results in a "Stage 2" maturity enterprise business include a number of common traits:

1. People get excited about the opportunity to make some changes and address issues.
2. The enterprise business wrestles with the difficulty of aligning improvement projects with the enterprise business planning and sustaining the gains from improvement projects.
3. Leadership wants it to work, but does not become personally engaged, hoping the employees it is been delegated to will follow through.
4. Therefore, the enterprise business ends up with isolated islands of improvement that fades over time. Improvements people wish to make often bump up against functional lines of authority and die a slow death. But some improvements do stick, and the enterprise business is getting better. Unfortunately, they are doing it at a slower pace than the rest of their industry, so they are slightly losing ground from a competitive perspective.

2.2.2.3 Stage 3: Proactive: Tools Master

The third stage of maturity, "Structure & Proactive," characterizes enterprise businesses that have developed formal structures and processes within the enterprise to comprehensively and proactively engage in strategic planning and management activities. These key activities occur on a fairly regular basis and are subject to some degree of improvement over time. Performance measures are somewhat aligned with the enterprise intended strategy and employee accountability is taken seriously.

At this third stage, there is a well planned, systematic and foundational performance measurement and management effort. This basic stage enables enterprise businesses to take advantage of at least some of the functionality that performance measurement and management have to offer. However, in order for the enterprise businesses to tap into the real power of performance measurement and management, and process improvement and management, it is important to progress far beyond this basic stage.

At this third stage, an enterprise business knows the improvement tools very well. It has a number of employees who would be considered as Tool Masters (e.g., Black Belt, Master Black Belt in Lean Six-Sigma environments). Because of the tool emphasis, the improvement activities are run by a select group of Tool Masters. Those individuals may serve as project team leads or facilitators. Enterprise businesses at this level are slightly gaining ground on their overall industry. Here, the enterprise business tends to focus on improvement projects, so most improvement primarily happens through project team activities.

An enterprise business at "Stage 3" maturity is very likely to have a vice president or director of improvement who coordinates improvement activities. People inside such an enterprise business are probably not aware of this, but the way they are going about improvement is a very normal approach, and very similar to how most of their competitors are going about it.

At this third stage, several people inside the enterprise business know the improvement tools very well and could also be considered Tool Masters. A few people on the executive management team are more engaged in the improvement activities than in "Stage 2" maturity enterprise businesses. But the improvement structure exists in parallel to the line organization running the business, so the improvement process is not actually part of how the business is managed. There is also more of an operations focus. Marketing, sales and administrative activities are typically not as engaged in improvement as the operations portion of the business.

People feel very good about their accomplishments, but for the most part, the gains are still hard to fully sustain. Improvement efforts tend to focus on functional departments directly linked to operations. Business performance measures focus more on the functional silos rather than cross-functional process performance. Authority still primarily sits in functional silos. Therefore, enterprise businesses at this third stage of maturity largely try to do a better job of what they already do versus creating new cross-functional business capabilities.

Enterprise business leaders in a "Stage 3" maturity certainly want the improvement initiative to succeed, but executive management members are typically busy running the business, and they expect the improvement experts – that is, the parallel organization to deliver project savings.

When this occurs; that is, when the executive management members abdicate their improvement responsibility to a staff group, a new power structure is created in the enterprise business. The enterprise business staff will complain about the lack of executive management support, but they will take on more responsibility trying to find the "right" things to improve. It is a chicken and egg problematic situation in which a desired outcome or improvement solution is impossible to attain because of a set of inherently illogical rules or conditions set within the enterprise business by the executive management team behavior.

While the staff leaders are trying to do the right thing, they remove responsibilities from the line organization. The improvement staff takes more responsibility for shepherding what should be improved, rather than line managers. And the line managers are busy with the real business, so they let the staff take that responsibility. That cycle continues, with the line organization never effectively assimilating daily improvement responsibilities. The strengths that got them that far to the third stage of maturity, which are now used to excess, become a weakness that inhibits the enterprise business' ability to transform. Therefore, the improvement initiative will likely remain isolated, and the enterprise business effectively fenced in at a "Stage 3" maturity.

In monthly or quarterly meetings, executive management board review improvement projects as an independent subject, separate from conversations about how well the business is operating. They often have a hard time getting in touch with reality relative to their competitors and to true customer needs. They are too inwardly focused.

Enterprise businesses at this stage of maturity have implemented meaningful changes, and they are better than average in their industry. Unfortunately, leadership often believes that they are at a "Stage 4" of maturity because so much has improved in the way the enterprise business operates. This disconnects from reality, relative to the outside world, becomes a major roadblock and decreases the probability of real transformation; it could fence in the enterprise business in its "Stage 3" of maturity. We are all blinded (trapped) to a degree by the assumptions we make, beliefs we hold, and the limited number of data points we know about.

Furthermore, at this stage of maturity, millions in savings might have been reported, but cost savings in a "Stage 3" maturity enterprise business is a pretend world. Based on the reported "savings," a "Stage 3" maturity enterprise business may appear to have improved, but the key business performance measures (financial, market share, etc.) often do not show a significant change. A "Stage 3" maturity enterprise business might be doing better than 50 % of its respective industry, and people believe that process cost savings results reported from projects will be a seamless addition to the bottom line of the cash flow statement, but the reality is often very different. Half of the savings reported are most likely "soft savings," where time or capacity was made available. And although no one reports this, the feeling is that somehow these soft savings will automatically turn into hard dollar savings.

Unfortunately, soft savings only turn into hard savings when leadership proactively does something to make it happen (i.e., cut expenses or use the freed-up capacity to make and sell more products or services to the customers). By primarily focusing on cost savings, leadership is not taking a hard look at what needs to be done to grow the business and foster better relationships with customers. Those conversations probably do take place in the enterprise business, but they happen elsewhere, and not in conversations regarding improvement initiatives. This often separates the enterprise business' performance improvement initiative from the real business.

2.2.2.4 Stage 4: Managed and Focused: Transformation

The fourth stage of maturity, "Managed & Focused," characterizes enterprise businesses where the intended strategy drives focus and decision making for the enterprise. Organization-wide standards and methods are broadly implemented for strategy management. Enterprise business executives, managers, and leaders formally engage employees in critical activities and performance measures and accountability help drive strategic success for the enterprise.

Thus a "Stage 4" maturity enterprise business has begun to transform its business. Enterprise businesses at this stage of maturity have a pretty good knowledge of the basic improvement tool set, but more importantly, the entire enterprise business – not just the Tool Masters – is driven to serve customers and stay ahead of the competition. Leaders in "Stage 2" and "Stage 3" enterprise businesses strive to serve customers better, but a "Stage 4" maturity enterprise business is much more effective at doing this. Enterprise businesses at "Stage 4" maturity begin to create new businesses, entirely new value streams, and add value to their customers in meaningful, systemic new ways. This transition from "Stage 3" maturity average to "Stage 4" and "Stage 5" greatness is a major challenge.

2.2.2.5 Stage 5: Continuous Improvement: Winning Team Culture

The fifth stage of maturity, "Continuous Improvement," characterizes enterprise businesses where the strategic planning and management excellence are part of the daily activities within the enterprise and are continuously improved in a formal sense. This means that as the enterprise performance is evaluated, the enterprise analyzes how it is performing towards its intended strategic goals and assesses and adapts as necessary how effective the strategic planning and management processes are. Excellence in strategic management drives the enterprise competitive edge or performance success.

Thus, a "Stage 5" enterprise business has a culture different from any of the other four stages. "Stage 5" maturity enterprise businesses are also very rare (less than 5 % of an overall industry). Enterprise businesses at this level tend to focus less on how good they are and more on how good they are not. Employee engagement and commitment in a "Stage 5" maturity enterprise business is more than twice as high as a "Stage 3" maturity.

Leaders in a "Stage 5" maturity enterprise business are very much in touch with reality. People do not try to hide problems or resolve them quietly out of sight. Leaders are just as concerned about a near miss as they are an actual defect or error. These enterprise businesses are very process and outcome focused: they are constantly striving to create the "perfect" process for their main business activities. Perfection is their primary goal and they measure their progress in terms of how far short of perfection they fall. A "Stage 5" enterprise business is on the journey toward True North.

Enterprise businesses do not automatically progress through the five levels. A "Stage 2" maturity enterprise business can certainly mature from learning about the tools and become a "Stage 3" tool master. The important challenge is to move beyond "Stage 3" into the top of the industry, and that is the primary focus of this book. Most enterprise businesses never progress beyond "Stage 3" maturity. The better job an enterprise business does of being a tool master, the less likely it is to progress into the top of its industry. That tool strength, when used to excess, actually becomes a weakness that prevents the enterprise business from progressing to the next stage of maturity: in other words, the enterprise business gets trapped in "Stage 3."

2.3 How to Realize a State of "Continuous Improvement"?

Moving an enterprise business to a higher maturity stage is one of the highest leverage activities that any enterprise can perform. Too many businesses have tried, but have failed miserably to develop the capabilities to improve incrementally and on an ongoing basis by adopting short-term "process improvement programs" or changing superficial aspects of their enterprise structure, systems, or technology without handling all the determining factors that characterize the "Continuous Improvement" maturity stage.

Successful enterprise businesses have to balance two needs – (1) the need to look backward in order to maintain the existing business and its current customers and (2) the need to look forward in order to explore and achieve performance breakthroughs and to identify and attract new customers and new sources of value. Achieving this balance requires specific and intensive actions along eight overarching determining factors of strategic management, that matter the most, among dozens:

1. Leadership
2. Culture and Values
3. Strategic Planning and Management
4. Performance Measurement
5. Performance Management
6. Alignment and Commitment
7. Process Improvement and Management
8. Sustainability

As indicated already in the introduction chapter, these overarching determining factors of strategic management were designed by The Balanced Scorecard Institute in its Strategic Management Maturity Model (SMMM) to:

1. Help enterprise business leaders – from CEOs to supervisors – perform a quick assessment of where their enterprise business stands in terms of strategic management;
2. Monitor progress in improving maturity of their enterprise business; and,
3. Allow benchmarking across organizations, or departments within one organization, in order to identify best practices.

We focused on these eight overarching determining factors of strategic management because they are the ones that matter the most and they are frequently causes of business improvement stagnation and failure. Also, they are rarely covered in context with the popular continuous improvement methodologies that enterprise businesses so earnestly use to get maximum leverage for their journey to competitive advantage.

If these eight problematic determining factors of strategic management are not in place, most business performance improvement efforts will fail to significantly change an enterprise business' competitive position. However, once an enterprise business grasp and do these eight well, its journey will not end. The eight overarching determining factors of strategic management are only meaningful because most (average) enterprise businesses do not perform them well. If more enterprise businesses begin to address these eight overarching determining factors of strategic management, that ups the ante. Then, they will need to push the frontier working these determining factors at a more sophisticated level in order to further improve their competitive position.

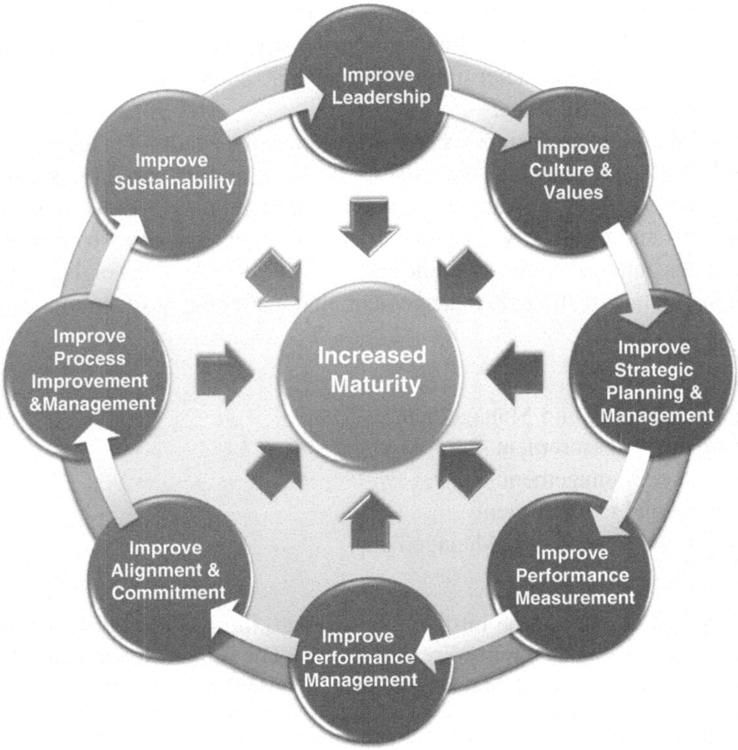

Fig. 2.4 Enterprise business maturity and dimensions' influence

Figure 2.4 depicts the influence of these determining factors – dimensions – on the enterprise maturity stage. Making progress toward the "Continuous Improvement" maturity stage requires improving simultaneously each of the eight determining factors. Thus the enterprise business maturity stage is a matter of both breadth and depth. Breadth indicates completeness of the eight overarching determining factors of strategic management. If any one these factors are weak or missing, then the results of the whole are jeopardized. Depth indicates the level to which the overarching determining factors of strategic management are performed, relative to a "world-class" standard.

Very few enterprise businesses manage to achieve performance of an overall maturity stage reaching "Stage 5" (i.e. "Continuous Improvement"): this is nearly 5 % of an industry. It is a challenging target. As more enterprise businesses begin to exhibit what used to be a "Stage 4" or "Stage 5" maturity stage, the leaders have moved on. They are not standing still. This is one of the reasons why some enterprise businesses, notably Toyota and GE, have been so hard to catch. Every year such enterprise businesses focus on how they can do a better job of getting better. It will keep an enterprise business humble, if it can maintain that mindset – something Toyota is currently trying to regain.

Improvement of these determining factors does not necessarily occur in a particular order, it is important that all eight determining factors be incrementally and simultaneously improved. While improvements occur along these determining factors, maturity (at the center of the diagram) increases. This maturation process is about movement to a higher stage of development, whereas improvement in each overarching determining factor is about "change in form, quality, or state, over time." Here, time is the "ether" of change and we judge that change has occurred against the background of time.

Not all changes result in improvements, thus we use metrics on the background of time for assessing when changes occur, the rate of change, and the extent of change, and also to establish the opposite of change, stability. It is the focus on change and an understanding of the basic principles of improvement that leads to efficient and effective improvement efforts in each overarching determining factor. Improvement has meaning only in terms of observation based on given performance measure.

In developing changes in form, quality, or state, over time, it is useful to distinguish between changes that are needed to sustain a determining factor at the current level of performance (reactive changes) and changes that are needed to create a new system of performance for the determining factor (fundamental changes).

Reactive changes in form, quality, or state, over time, are required to maintain a determining factor (or its constituents) at its current level of performance. Here are some aspects of reactive change:

1. They are often made routinely in reaction to a special circumstance, which may also affect other parts of an enterprise business.
2. They often result in putting the determining factor (or its constituents) back to where it was some time before.
3. They typically take the form of tradeoffs among competing interests or characteristics (such as increasing quality but also increasing cost, reducing errors but also reducing volume, or speeding up delivery but reducing customer service).
4. Their impact is usually felt immediately or in the near future.

When one faces a problem, making a reactive change in form, quality, or state, over time is often the best approach to immediately solving the problem and restore the performance of the system of interest to its previous level. The ability to make reactive changes in form, quality, or state, over time is very important for any enterprise business.

Reactive changes should not, however, be confused with fundamental changes. Fundamental changes in form, quality, or state, over time are required to improve a determining factor (or its constituents) beyond historical levels. Here are some important aspects of fundamental change:

1. They result from design or redesign of some aspect of the system (process, product, or service) or the system as a whole.
2. They are necessary for the improvement of a system that is not plagued by problems.

3. They fundamentally alter how the system works, what people do and how people perceive the system.
4. They often result in improvement of several aspects simultaneously (quality and cost, or time-to-market and errors).
5. Their impact is felt into the future.

A change in form, quality, or state, over time is characterized by its characteristic rate, rhythm, or pattern of work or activity (Weick & Quinn, 1999). It can be episodic or continuous. Episodic change in form, quality, or state, over time is conceived to be "infrequent, discontinuous and intentional," while continuous change in form, quality, or state, over time is conceived as "ongoing, evolving and cumulative."

The two forms of change in form, quality, or state, over time are associated with different metaphors of the enterprise business, analytical frameworks, conjectures of intervention, and roles attributed to change agents, as shown in Table 2.1. The distinction between episodic and continuous change in form, quality, or state, over time is correlated with several others, including incremental versus radical change continuous versus discontinuous change, first-order versus second order change and competence enhancing versus competence-destroying change.

Improvement in each determining factor may incorporate elements of both continuous and episodic change in form, quality, or state, over time. Continuous change in form, quality, or state, over time generally occurs at the micro level of system behavior and concrete actions that move processes through (episodic) stages.

Whereas stage change in form, quality, or state, over time is conceptualized as episodic. Stage wise change presumes an underlying continuous process of activity as a means for constructing the stages. Thus, episodic change is best understood from a macro or global analysis, while continuous change is better discerned through micro level or local analysis.

As we consider the enterprise maturity, from the perspective of systems theory, it is important to remember that "Continuous Improvement" maturity is not about strength in any one aspect of the enterprise, but about the health of the enterprise as a "total system."

Some enterprise businesses excel at one, two or three of the eight determining factors. However, for any enterprise to achieve superior results, it is essential that all eight determining factors work in tandem with each other.

For example, without the right leadership, the other factors will be meaningless – because if you do not influence people and to gain their genuine commitment to accomplish common organizational goals, you will not be able to create the right culture, you will not be able to plan and manage strategically, you will not be able to improve the enterprise performance, you will not be able to improve internal processes, and you will not get the right results.

On the other hand, even with the right leadership, without positive culture and values, people won't be motivated to improve the right things, employees will tend to focus on what will bring them the largest personal rewards, and will tend to have an adversarial posture toward whatever it is that the enterprise initiates.

Without the right alignment and engagement, improvement initiatives will stand alone, individuals and functions will not be properly aligned, and there will be a natural tendency to maximize individual gains, often at the expense of other parts of the enterprise. Without frequent interaction relative to performance measurement,

Table 2.1 Comparison of episodic and continuous change

Characteristic	Episodic change	Continuous change
Metaphor of enterprise business	Enterprise businesses are inertia-prone and change in form, quality, or state, over time is infrequent, discontinuous, and intentional	Enterprise businesses are emergent and self-organizing and change in form, quality, or state, over time is constant, evolving, and cumulative
Analytic framework	Change in form, quality, or state, over time is an occasional interruption or divergence from equilibrium. It is externally driven. It is seen as a failure of the enterprise business to adapt to a changing environment	Change in form, quality, or state, over time is a pattern of endless modifications in work processes and social practice. It is driven by organizational instability and alert reactions to daily contingencies. Numerous small accommodations cumulate and multiply
	Perspective: Macro, distant, global	Perspective: Micro, close, local
	Emphasis: Short-run adaptation	Emphasis: Long-run adaptability
	Key concepts: Inertia, deep structure, or interrelated parts, triggering, replacement and substitution, discontinuity, revolution	Key concepts: recurrent interactions, response repertoires, emergent patterns, improvisation, translation, learning
Conjectures of intervention	Intentional change: Unfreeze, change, and refreeze	Redirection of existing tendencies
	Change in form, quality, or state, over time is inertial, linear, progressive, and requires outside intervention	Change is cyclical, process based, without an end state, equilibrium-seeking, eternal
Role of change agent	Prime mover who creates change in form, quality, or state, over time by finding points of leverage in the enterprise business	Sense maker who redirects and shapes change in form, quality, or state, over time
	Change agent changes meaning systems, schema, and punctuation	Change agent recognizes, makes salient, and reframes current patterns. Change agent unblocks improvisation, translation, and learning

performance management, and strategy, none of the other determining factors can really function.

When all eight determining factors of strategic management are working together synergistically, the creative energy within the enterprise is release to make a real difference – a transformational difference – in your enterprise!

2.4 Where Does Your Business Fit?

The first step that enterprise business executives and managers must take for a "Continuous Improvement" transformation program implementation must be to assess their current enterprise maturity stage by scoring their stage of performance on each of the five maturity stages, and for each of these eight dimensions.

In assessing the enterprise current maturity stage, a review and analysis of the enterprise practices and behaviors must be conducted in sufficient details and benchmarked against the models descriptions illustrated in Table 2.2 adapted from the Balanced Scorecard Institute online publication on the Strategic Management Maturity Model (SMMM).

A quick assessment can be made by observing the practices and behaviors within the enterprise. To do this, scan the five maturity stages from this table, then locate and circle the characteristics that best describe your enterprise. A pattern should begin to emerge, identifying your enterprise's maturity stage, primarily in a single stage. All your circles may not be confined to one stage, however, because some business functions will be more developed than others.

Keep in mind that if you rate your effectiveness at a "Stage 4" or "Stage 5" maturity, you are claiming there are things your enterprise business could teach to the best companies in the world – organizations as sophisticated as Toyota or GE – about how to more effectively improve for that particular trait. The purpose of this quick assessment is not to see how high of a number your enterprise business can score. The assessment is intended to assist your enterprise business in having an open dialogue about what is important to improve.

After having identified your enterprise current maturity stage, use the next maturity stage as a vision to focus on the improvement concerns that you are facing now – and may face at the next maturity stage.

It may be helpful to display the result on a radar graph as illustrated in Fig. 2.5. This is necessary to attract management's attention and interest on the current status of maturity of the business and to set a vision and objectives for closing the required horizontal (maturity) and vertical (dimensions) gaps.

The identified horizontal (maturity) and vertical (dimension) gaps are of course opportunities for business improvement. The identification of maturity gaps can also be used to evaluate the quality of the enterprise's forward strategic and business planning.

The continuous discovery of maturity gaps may indicate that unrealistic targets are being set, or that the assumptions and theories upon which these targets are being based are incorrect. These gaps can also reveal something about the enterprise's capacity for effective strategic planning and management, as much as about its operational performance. The specific methodology to be used to close the determined gaps is be described in our next volume.

Before getting to the next chapters of the present book, let's mention that when people rate their performance, there is a bias toward a higher self-rating than the actual reality. Most of us believe we work hard, we believe we do the right thing, and mostly we are simply not capable (at the beginning) of seeing how much opportunity for improvement exists, all around us. So when you look at the instrument, focus mostly on the descriptions given in Table 2.2. Which description most fairly describes your enterprise business reality? And how can you do a reality check?

In the following chapters, we will look at the essentials of each of these eight overarching determining factors of strategic management with the purpose of giving both:

Table 2.2 Enterprise business maturity stages and dimensions assessment

	Stage 1: Ad hoc and static	Stage 2: Reactive	Stage 3: Structured and proactive	Stage 4: Managed and focused	Stage 5: Continuous improvement
Leadership	Leader dictate, command and control; otherwise disengaged	Leaders dictate but gather feedback sporadically	Leaders engage with direct reports only, but no model desired behaviors and values	Leaders empower many employees through ongoing engagement	Leaders and employees fully engage in a continuousdialog based on a team-based culture
Cultures and values	Vision & Values undefined or not shared	Vision & Values published, but not lived	Vision & Values communicated and understood	Vision & Values collaboratively developed	Vision & Values fully integrated into enterprise culture
Strategic planning and management	No strategic planning occur within the enterprise; no goals defined	Strategic planning is the responsibility of a small team and dedicated to the enterprise	A structured and open planning process involves people throughout the enterprise every couple of years	Plans are developed and revised regularly by trained, cross-functional planning teams	Strategy drives critical enterprise decisions and a continuous improvement planning process is maintained
Performance measurement	No data, or only ad hoc performance measures are collected	Performance data collected routinely, but mostly operationally focused	Strategic performance measures are collected, covering most strategic objectives	Strategic measures are broadly used to improve focus and performance and inform budget decisions	Measurements comprehensively used and routinely revised based on continuous improvement
Performance management	No emphasis on using performance as a criterion to manage the enterprise	Performance reviews required but not taken seriously; no accountability for performance exists	Measures are assigned owners and performance is managed at the organizational and employee levels	Measurement owners are held accountable and performance is managed at all levels	Enterprise culture is measured and accountability focused; decisions are fact based
Alignment and commitment	Work is narrowly focused based on enterprise structure, with little customer input	Customer needs and feedback start to influence more aligned decision making	Employees know their customers and align strategy to those needs	Vision, customer needs, strategy, and employee reward and recognition systems are cascaded and aligned	All structures and systems are aligned with strategy, and organizational alignment is continuously improve
Process improvement and management	Processes are undocumented and ad hoc with evident duplication and delays	A few key processes documented, and process improvement model introduced	All key processes are identified and documented, and strategy guides successful process improvement initiatives	All key processes are tracked and improved on a continuous basis and new process improvement ideas are accepted	Employees are empowered and trained, and a formal process exists for improving process management
Sustainability	Lack of structure and champions lead to short-term focus on tasks	Strategy champions identified	Formal enterprise structure in place to maintain focus on strategy	Enterprise has an "Office of Strategy Management" or equivalent	Strategic thinking and management are embedded in the culture of the enterprise

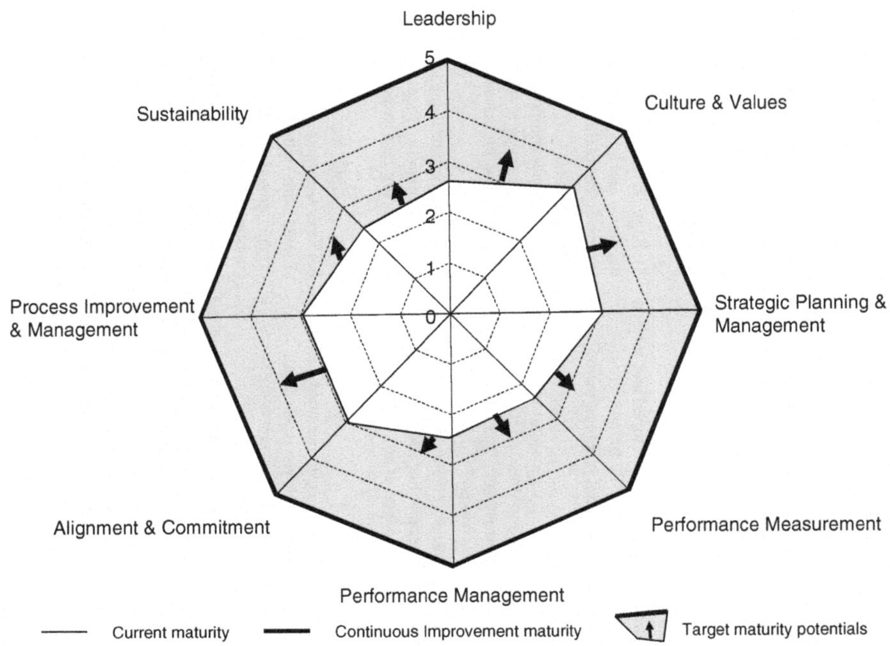

Fig. 2.5 Current and target maturity potentials

1. The necessary understanding to our readers' enterprise business leaders – Project Managers, Green Belts, Black Belts, managers at all levels, and process improvement professionals – on the issues that matter most to achieve transformation, survival and success; and
2. A reality check for your enterprise business capability to progress toward a "Continuous Improvement" maturity stage.

3 Understanding Leadership Dimension

A reality accepted by most professionals engage in any "Continuous Improvement" program implementation, but rarely understood enough to be described accurately is "leadership."

Over time and in media, the idea of leadership has to grown to being synonymous with executive management at best and as just another skill or trait that makes up the competent manager at worst. Furthermore, when people talk about "developing leaders" they mean developing prospective executive managers. When they ask, "What do the leaders think about this initiative?" they are asking about the views of the executive managers. As Peter Senge pointed out (Senge, 1994), there are two problems with this conception.

1. First, it implies that those who are not in executive management positions are not leaders. They might aspire to "become" leaders, but they do not "get there" until they reach a senior management position of authority.
2. Second, it leaves us with no real definition of leadership. If leadership is simply a position in the hierarchy, then, in effect, there is no independent definition of leadership. A person is either an executive manager or is not. There is nothing more to say about leadership. End of the story. To understand the correct significance of leadership, we must explicitly determine the difference between management and leadership. Failing to make and understand this difference could undermine a successful implementation of a "Continuous Improvement" program.

This chapter delves into the key characteristics and constituents of the leadership dimension necessary to take your enterprise business to the "Continuous Improvement" stage of maturity as described in the previous chapter.

3.1 What Is Leadership?

Enterprise businesses depend upon people who keep the business activities moving along, ensure productivity, and control and schedule the use of appropriate resources, but enterprise businesses also need people who can infuse the business

A. Van Aartsengel and S. Kurtoglu, *A Guide to Continuous Improvement Transformation*, Management for Professionals, DOI 10.1007/978-3-642-35904-0_3,

with purpose and values, help determine the character of the enterprise, and ensure its long-term survival. The skills and competencies required to achieve the first critical activity are substantially different than those needed to achieve the second. The first is the domain of the manager; the second is the domain of the leader. Although most professionals and practitioners engage in any "Continuous Improvement" program implementation continue to confuse the two concepts or make no distinction.

For most people, leadership, or a position of leadership, is a management role with its accompanying tasks and techniques – its technology of control, making decisions, aligning and setting corporate goals, planning, budgeting and directing the effort of the several followers engaged in work. The manager role entails insuring that group activity is timed, programmed, controlled, and predictable. Management is the science or art of achieving goals through people. Whatever the title given, be it top manager, middle manager, director, team or project leader, or department head, etc., the task remains constant: "*to get the job done and keep the people motivated while operating within numerous restrictions such as time, quality, costs, limited resources, rules, and tradition.*"

We view leadership as "*the art of influencing people and to gain their 'genuine commitment' to accomplish common organizational goals,*" while management is "*the science of specifying and implementing means needed to accomplish these ends.*"

In today's work environment, managers have fewer tools to influence employee behavior because coercion is no longer an option. No intelligent manager would hope to obtain in any full measure the "genuine commitment" of people unless he/she felt that he/she was giving them something more than they usually receive from the enterprise business. It is well within the mark to affirm that:

> *In almost all enterprise businesses, most employees believe it to be directly against their interests to give their employers their 'genuine commitment,' and that instead of working hard to do the largest possible amount of work and the best quality of work for their enterprise business, they deliberately work as slowly as they have the courage while at the same time try to make those over them believe that they are working fast.*

In order to have any hope of obtaining "genuine commitment" of people, and help the enterprise business move to the "Continuous Improvement" maturity stage, managers must endeavor to create positive work environments that can influence, but not order or command, desired behavioral outcomes.

In most enterprises that have not yet reached the "Continuous Improvement" maturity stage, efforts to increase income, lower expenses, and maximize profit in the short term are built around employees "*compliance*" – *the forced adherence to plans created through manipulation, punishment, and coercion.* They do not require "*commitment – the innate willingness of people to follow and contribute.*" Either people comply with the instructions, or they know they will be at odds with their manager. Knowing that it is difficult to discern visions from commands when they travel down the hierarchy, savvy senior managers use the power of their position – because they seek to foster more than just compliance.

For "Continuous Improvement" transformation to occur, you as enterprise business executive and manager must realize that your power is in fact limited and lasting transitions and changes – in how individuals think within the enterprise business, what they believe, how they see the enterprise business – are difficult, if not impossible, to achieve through compliance. Individuals cannot be expected to comply with the instructions to increase income, lower expenses, and maximize profit if they are unable to find value in that. They do not change their behavior without knowing what new behavior is expected of them and why. They cannot make decisions unless they have information; then they will make the best decisions they can with the amount of information available to them. Thus little significant transition from the current enterprise business maturity stage to the "Continuous Improvement" maturity stage can occur if it is driven only from the managers. This does not imply that effective management is unimportant.

Making every individual, system, activity, program, and policy countable, measurable, predictable, and therefore controllable are realities of management. While these emphases are important in managing things, they are not used as basis for leading employees within enterprises at the "Continuous Improvement" stage of maturity. Leadership subscribes to a different reality than management and we must look at it differently.

3.2 Leadership Characteristics

Leadership is a transforming values-laden and relationship-based process that occurs in reciprocal relations within a group of people. No one person has sole responsibility for leadership within a group of people. Leadership is provided by anyone who helps create and maintain the performance-enhancing conditions within a group of people, hence within an enterprise business, regardless of whether that person happens to hold a formal leader role.

As a transforming process, leadership implies changing the individuals within a group as well as the group to enable these individual to reach higher stages of accomplishment and self motivation. It releases human potential for the collective pursuit of common goals. This is done by fostering, through business activities, an environment where people have freedom of thought, are comfortable talking about their different values and aspirations for the enterprise as a whole, and can take action to realize their values-laden vision with no fear of persecution or retribution.

Although employees within an enterprise business may isolate some specific definitional elements of leadership, these elements may not be understood fully nor put into practice at all except through the individual's conception and perspectives.

3.3 Leadership Perspectives

The notion that values play a key role in enterprises provides a way to frame the variety of individual perspectives about values, enterprise business, and leadership. An elaboration of these perceptions is described by Fairholm's Leadership Perspectives Model (LPM), which proposes an interrelated hierarchy of individual's conception about what leadership is (Fairholm & Fairholm, 2008).

The model, illustrated in Fig. 3.1, emerged in part as a result of studying the attitudes and values of practicing organizational leaders, in part from analysis of available past and contemporary literature, in part from observation of leaders in action, and in part from the authors' personal experiences. It is soundly based on both research and practice and is useful, as theory should be, for both descriptive and prescriptive purposes. It is descriptive in the sense of exploring how one may perceive leadership and positioning that perspective into an overarching leadership model.

Fairholm's Leadership Perspectives Model explains the activities, tools, approaches, and techniques required to be effective or successful within each of the five perspectives. The five perspectives, themselves, are legitimate constructs that aid understanding about how individuals may view leadership and together outline a comprehensive leadership model.

The first two perspectives focus on values that relate to organizational hierarchy and authority within the enterprise. The last three take into account a more personal approach to values. Each perspective had its period of prominence in the past. Each is true in that it helps describe some part(s) of the leadership task. They each lay out a logical, rational – although incomplete – pattern of leadership actions. It is only together that they define the full picture of leadership.

Below are brief descriptions of the different stages of leadership that form the basis for Fairholm's Leadership Perspectives Model.

1. *Leadership as (Scientific) Management* – Leadership equals management in that it focuses on getting others to do work the leader wants done, essentially separating the planning (management) from the doing (labor).
2. *Leadership as Excellence Management* – Leadership emphasizes quality and productivity process improvement rather than just product and people over either product or process, and requires the management of values, attitudes and organizational aims within a framework of quality improvement.
3. *Values Leadership* – Leadership is the integration of group behavior and shared values through setting values and teaching them to followers through an articulated vision that leads to excellent products and service, mutual growth, and enhanced self-determination.
4. *Trust Culture Leadership* – Leadership is a process of building trust cultures within which leader and follower (in an essentially voluntary relationship, perhaps from a variety of individual cultural contexts) trust each other to accomplish mutually valued goals using agreed-upon processes.
5. *Spiritual (Whole-Soul) Leadership* – Leadership is the integration of the components of work and self – of the leader and each follower – into a

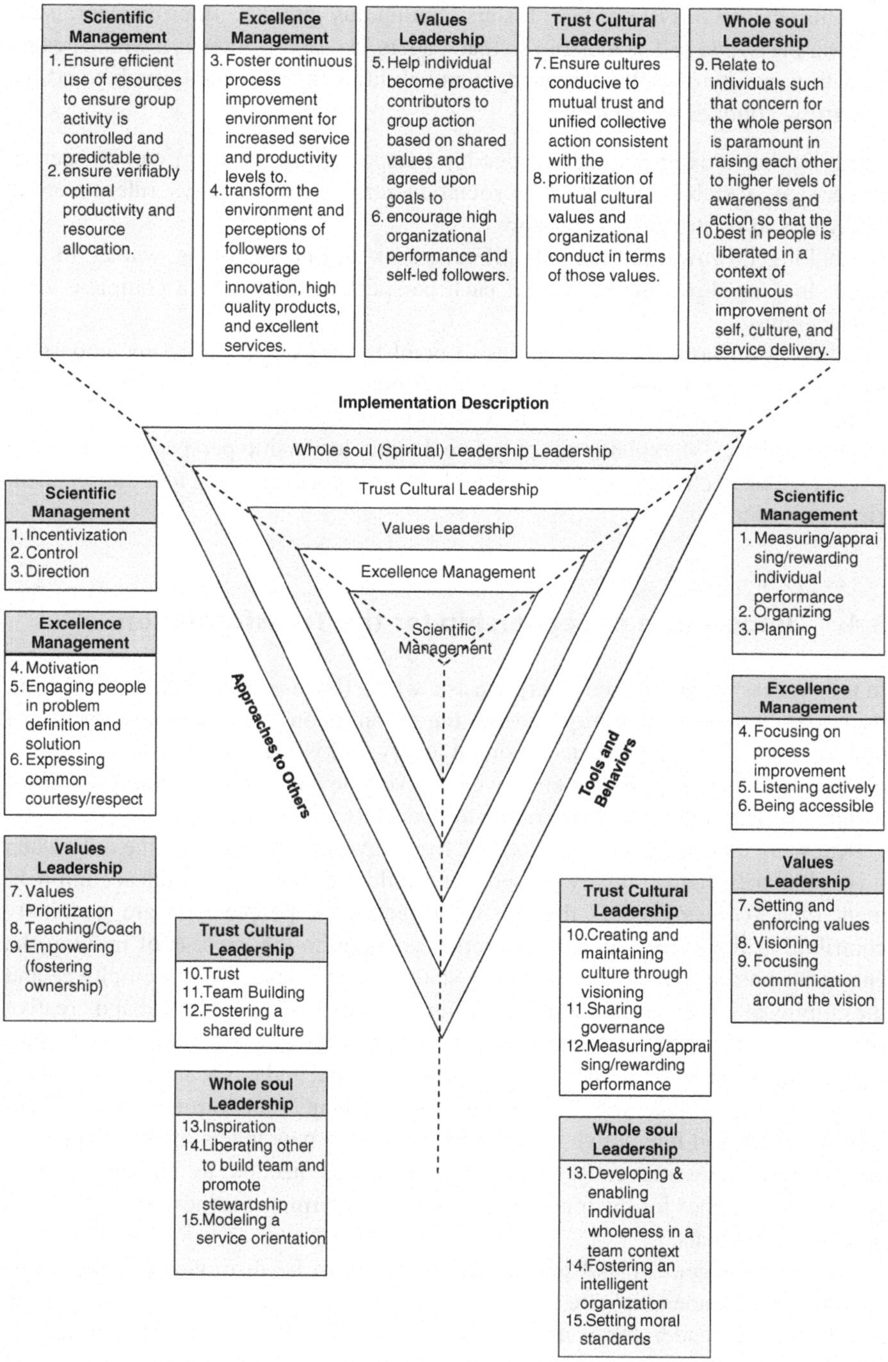

Fig. 3.1 Leadership perspective model (Adapted from (Fairholm & Fairholm, 2008))

comprehensive system that fosters continuous growth, improvement, self-awareness, and self-leadership so that leaders see each worker as a whole person with a variety of skills, knowledge, and abilities that invariably go beyond the narrow confines of job needs.

As Fairholm explains, each leadership perspective in the LPM can be seen in essence as a stable portion of a social hierarchy that displays rule-governed behavior and/or structural constancy.

Looking downward, each subunit serves as an encompassing whole. In this sense, looking down the hierarchy, each perspective looks like a complete view of leadership.

Looking upward, each unit serves to point toward larger, more encompassing ways of engaging or understanding the whole notion. In this sense, each perspective points to and grounds a broader, more holistic view of leadership.

Hence, the LPM explains not only five distinct leadership perspectives, but also how each perspective builds toward a higher, more encompassing and transcendent view of leadership.

3.4 Importance of Leadership for the Transformation

In the process of moving the enterprise as a whole (businesses and customers) from its current stage of maturity toward the "Continuous Improvement" stage of maturity the collective participation of every individual at each levels of the enterprise is required. Every employee at every level within an enterprise must assume leadership role and contribute to the effort.

Knowing that "*genuine commitment*" from any individual is of value only when it is voluntarily and genuinely chosen, we should see leadership as the accomplishment of a "*common goal*" through the direction of people who are genuinely contributing their creative and productive energies to the process of moving the enterprise business to a higher maturity state. Commitment of every employee and the employee involvement should be limited only by his/her analytical and creative capability, and not by his/her position level on the enterprise organizational chart. Leadership as we have defined in the previous section, and advocate in this book, is needed not just to make the "Continuous Improvement" transformation contextualized, focused, and interactive – and so productive at new levels of effectiveness – but to apply systematically the critical resources needed to realize the rich potentials describes for the transformation of the enterprise business and empowerment of individuals.

Just as management and leadership are terms to be distinguished, the terms "leader" and "leadership" are also not synonymous, nor are they interchangeable. A values-laden leader, for example, fosters a positive context environment; that is, an environment where people have freedom of thought, are comfortable talking about their different values and aspirations, and can take action to realize their values-laden vision with no fear of persecution or retribution.

In today's business world, change is chaotic and unpredictable. In such a world, leadership is at a premium and takes precedence over management. Leadership is most needed when enterprise businesses must change direction or style, when they must shift, adapt, or move to respond to changing circumstances. Management, on the other hand, is concerned with production, consistency, and flow. It is most needed in times of stability and predictability.

The leader's authenticity is primordial as he tries to impact enterprise business dynamics such as creativity, relationships, and innovation and attempt to create trusting work environments. Inspired leaders give voice to other, assist them, listen to them, and positively impact their lives. Leaders think differently, value things differently, and relate to others differently. They infuse the group with values, have their own unique expectations for others and seek different results from individuals and from the group than do managers. They impact stakeholder groups in volitional ways, not through formal authority mechanisms.

Culture and Values Dimension

4

There have been many definitions of an enterprise business or organization culture in recent years and most of these definitions refer to behaviors, shared values, beliefs, assumptions, and patterns of relationships (Alvesson, 2002; Ashkanasy, Wilderom, & Peterson, 2010; Brenton & Driskill, 2010; Cameron & Quinn, 2011; Deal & Kennedy, 2000; Kotter & Heskett, 1992; Mann, 2005; Martin, 2002; Parker, 1999; Pheysey, 1993; Schein, 2009, 2010; Witte & Muijen, 2000).

In the last several decades the terms "enterprise business culture" or "organization culture" have been used by many professionals to refer to the climate and practices that enterprises develop around their handling of people, or to the espoused values and credo of an enterprise. In such context, enterprise business executives and managers speak of developing the "right type of culture," "culture of quality," "culture of customer service," "culture of continuous improvement," "Lean culture" or "Six-Sigma culture," signifying that an enterprise business culture has to do with certain values that managers and executives are trying to inculcate in their business.

Regardless of how it is defined, enterprise culture and values are realities, often assumed and implicit rather than explicit, that affects what an enterprise business can and cannot do.

This chapter delves into the key characteristics and constituents of an enterprise business values and culture necessary to take your enterprise business to the "Continuous Improvement" stage of maturity as described in a previous chapter.

4.1 Enterprise Business Values

Values define the nature and character of the enterprise business. Values do not refer solely to what an enterprise business deems to be ethically acceptable. They define the standards by which employees set priorities that enable them to judge whether an order is attractive or unattractive, whether a customer is more important or less important, whether an idea for a new product or a new service is attractive or marginal, and so on. But consistent, broadly understood enterprise business values also define what an enterprise business cannot do.

A. Van Aartsengel and S. Kurtoglu, *A Guide to Continuous Improvement Transformation*, Management for Professionals, DOI 10.1007/978-3-642-35904-0_4,

An enterprise business values are the criteria its employees use when making prioritization decisions. Every employee at every level of the enterprise business makes these types of decisions every day. For example salespeople have to decide whether they should call customer A or customer B. Once they decide whom to call, they must make an on-the-spot decision about which products or services to push versus their priorities. Engineers decide which project they will work on today and which project they will tackle tomorrow. They make the design choice A and not the design choice B.

Values also determine the larger strategic decisions that senior management makes. Do we acquire this company or that company? Do we grant the budget request of this business unit to launch a "Continuous Improvement" initiative or cut the budget of that one? Values are the criteria that drive an enterprise business' resources-allocation process – the mechanism that defines which threats and opportunities the enterprise business will pursue, and which it will not.

An enterprise business values reflect its cost structure or its business model because those define the rules its employees must follow for the business to prosper. If, for example, an enterprise business' overhead costs require it to achieve gross profit margins of 30 %, then a decision rule will have evolved that encourages middle managers to kill ideas that promise gross margins below 30 %. Such an enterprise business would be incapable of commercializing projects targeting low-margin markets – such as those in e-commerce – even though another enterprise culture and values, driven by a very different cost structure, might facilitate the success of the same project.

Within an enterprise, values establish the foundation for specific operational and interpersonal work standards used by employees. A common set of values binds people together, while conflicting values disrupt and may even destroy a corporation or other work teams. Different enterprise businesses, of course, embody different values. But we want to focus on two sets of values in particular that tend to evolve in most companies in very predictable ways. The inexorable evolution of these two values is what makes enterprise businesses progressively less capable of addressing movement to higher states of maturity successfully.

As in the example here above, the first value dictates the way the enterprise business judges acceptable gross margins. As enterprise businesses add features and functions to their products and services, trying to capture more attractive customers in premium tiers of their markets, they often add overhead cost. As a result, gross margins that were once attractive become unattractive.

For instance, an automobile enterprise business, let call it "ABC Automobile Inc.", entered the market with its lead model, which targeted the lower end of the market. As that segment became crowded with look-alike models from its competitors, competition drove down profit margins. To improve its margins, "ABC Automobile Inc." then developed more sophisticated cars targeted at higher tiers. The process of developing cars like the competition added costs to "ABC Automobile Inc." operation. It subsequently decided to exit the lower end of the market; the margins had become unacceptable because the "ABC Automobile Inc." cost structure, and consequently its values, had changed.

In a departure from that pattern, "ABC Automobile Inc." recently introduced a new model, hoping to rejoin the entry-level tier with a lower car price. It is one thing for "ABC Automobile Inc." senior management to decide to launch this new model. It is another for the many people in "ABC Automobile Inc." system – including its dealers – to agree that selling more cars at lower margins is a better way to boost profits and equity values than selling more its original model, look-alike from its competitors. Only time will tell whether "ABC Automobile Inc." can manage this down-market move. To be successful with the new model, "ABC Automobile Inc." management will have to swim against a very strong current – the current of its own corporate values.

The second value relates to how big a business opportunity has to be before it can be interesting. Because an enterprise business' stock price represents the discounted present value of its projected earnings stream, most managers feel compelled not just to maintain growth but to maintain a constant rate of growth. For a 50 million EUR enterprise business to grow 30 %, for instance, it needs to find 10 million EUR in new business the next year. But a 50 billion EUR enterprise business needs to find 10 billion EUR in new business the next year to grow at that same rate. It follows that an opportunity that excites a small enterprise business is not big enough to be interesting to a large enterprise business. One of the bittersweet results of success, in fact, is that as enterprise businesses become large, they lose the ability to enter small, emerging markets. This disability is not caused by a change in the resources within enterprise businesses – their resources typically are vast. Rather, it is caused by an evolution in values.

Thus, the values of successful enterprise businesses tend to evolve in a predictable fashion on at least two dimensions. The first relates to acceptable gross margins. As enterprise businesses add features and functionality to their products and services in an effort to capture more attractive customers in premium tiers of their markets, they often add overhead cost. As a result, gross margins that at one point were quite attractive at a later point seem unattractive. Their values change. The second dimension along which values can change relates to how big a customer or market has to be, in order to be interesting. Because an enterprise business' price represents the discounted present value of its projected earnings stream, most enterprise business executives and managers typically feel compelled not just to maintain growth, but to maintain a constant rate of growth.

4.1.1 Identifying an Enterprise Business Values

An enterprise business values can be identified by looking for proxies. An enterprise business' revenue mix, cost structure, absolute size, most important customers, and history of past investment decisions can help you understand the type of strategies and investments that will appear profitable to its managers, and which will appear unattractive. It is a pretty safe bet that enterprise business executives and managers will accord highest priority to the opportunities that are financially most attractive.

4.1.1.1 Income Statement

The first place to look to identify an enterprise business' values is its income statement. What is its revenue mix? Does it earn a significant proportion of its revenue from many products or a specific subset of products? From post sales services? An enterprise business is unlikely to prioritize opportunities that destroy significant revenue streams. What are the gross profit margins it needs to earn to support its cost structure? An enterprise business that has a cost structure that requires a 60 % gross margin is typically not interested in opportunities with a 20 % gross margin. How big does a new opportunity have to be to matter to the enterprise business? An opportunity that is attractive to an enterprise business that has EUR 60 millions in sale and seeks 10 % top-line growth would be unattractive to an enterprise business that less than EUR 5 billion in sales and seeks 10 % top-line growth. A EUR 2.5 million opportunity meets 50 % of the first enterprise business' growth needs but only 0.5 % of the second enterprise business' growth needs. Which enterprise business do you think will place a higher priority on going after a EUR 2.5 million market?

4.1.1.2 Customers

The next clue to an enterprise business' values comes from its customers' roster. An enterprise business needs to prioritize opportunities that improve its ability to serve its most important customers. Enterprise businesses that derive a high proportion of their income from a certain class of customers are likely to focus on opportunities that target those customers.

4.1.1.3 History of Past Investments

The final way to see an enterprise business' values is to look at its history of past investment decisions. Which opportunities did it decided to target and which opportunities did it decided to forgo? The streams of incremental investments decisions taken by an enterprise business during a past period of time indicate its focus values.

4.2 Understanding the Resources-Processes-Values Framework

Three classes of factors affect what an enterprise business can and cannot do: its resources, its processes, and its values. This concept, which originates from Clayton Christensen, a Harvard Professor of Business Studies, helps to capture the essence of an enterprise business culture (Christensen & Overdorf, 2010; Christensen, Anthony, & Roth, 2004).

4.2.1 Resources

In the start-up stages of an enterprise business, much of what gets done is attributable to resources, both tangible (people, equipment, technologies, cash flow) and intangible (product designs, information, brands, current relationships with suppliers, distributors and customers) – but mainly people, in particular. Resources are the most visible of the factors that contribute to what an enterprise business can and cannot do. They can be hired and fired, bought and sold, depreciated or enhanced. Resources are not only valuable, they are flexible. An engineer who works productively for an automobile factory can also work productively in a start-up enterprise business. Software that helps an automobile factory manage its logistics system can also be useful in a different industry. Technology that proves valuable in mainframe computers also can be used in telecommunications switches. Cash flow is a consummately flexible resource. The addition or departure of a few critical resources can profoundly influence an enterprise business success.

4.2.2 Processes and Values

Over time, however, the locus of the enterprise business' capabilities shifts toward its processes and values. As people address recurrent tasks, processes, which describe how the people who the enterprise business employs do interact, coordinate, communicate and make decisions, become defined. And as the business model takes shape and it becomes clear which types of business need to be accorded highest priority, values, which are the standards by which employees set priorities in making their decisions, both large and small, coalesce.

Prioritization decisions are made by employees at every level through processes. Among salespeople, they consist of on-the-spot, day-to-day decisions about which products to push with customers and which to de-emphasize. At the executive tiers, they often take the form of decisions to invest, or not, in new products, services, and processes. In fact, every organization and group that endures for even a modestly short time develops a culture and values, and each culture is unique.

When the enterprise business' processes and values are being formed in its early and middle years, the founder typically has a profound impact. The founder usually has strong opinions about how employees should do their work and what the enterprise business' priorities need to be. If the founder's judgments are flawed, of course, the enterprise business will likely fail. But if they are sound, employees will experience for themselves and trust the validity of the founder's problem-solving and decision-making methods. Thus processes become defined. Likewise, if the enterprise business becomes financially successful by allocating resources according to criteria that reflect the founder's priorities, the enterprise business' values coalesce around those criteria.

As the successful enterprise business matures, employees gradually come to assume and trust that the processes and priorities they have used so successfully so often are the right way to perform their work and daily tasks. Once that happens and employees begin to follow processes and decide priorities by assumption rather than by conscious choice; that trust in those processes and values come to constitute the foundation of the enterprise business culture.

As the enterprise business grows from a few employees to hundreds and thousands of them, the challenge of getting all employees to agree on what needs to be done and how can be daunting for even the best managers. Thus, trust in those processes and values that the enterprise business has used so successfully so often, which come to constitute the basis of enterprise business culture, is a powerful management tool in those situations. It enables employees to act autonomously but causes them to act consistently.

Accordingly, the factors that define an enterprise business' capabilities and disabilities evolve over time in a Resources-Processes-Values framework – they start in resources; then move to visible, articulated processes and values; and migrate finally to culture. As long as the enterprise business continues to face the same sorts of problems that its processes and values were designed to address, managing the enterprise business can be straightforward. But because those factors also define what the enterprise business cannot do, they constitute disabilities when the problems facing the enterprise business change fundamentally.

When the enterprise business' capabilities reside primarily in its people, changing capabilities to address the new problems is relatively simple. But when the capabilities have come to reside in processes and values, and especially when they have become embedded in trust in those visible, articulated processes and values come to constitute the enterprise business culture, change can be extraordinarily difficult.

4.3 Defining an Enterprise Business Culture

We can look at the "enterprise culture" from two viewpoints. Firstly, as founded in trust in those processes and values that the enterprise business has used so successfully so often that they give direction to its daily life. These processes and values prescribe the general ways employees relate to each other, whether in trusting or distrusting ways. This is a strategic, global perspective of an enterprise business culture, which proceeds from both internal and external guiding trust in processes and values. Secondly, as founded in the daily routine of an enterprise business through the accepted system of meanings that give direction to specific routine acts that each employee performs daily.

As a result, we can say that an enterprise business culture basically spring from three sources:

1. Trust in those processes and values that the enterprise business has used so successfully so often that they give direction to its daily life;

2. The learning experiences of employees and group members as the enterprise business evolves; and
3. New beliefs, values, and assumptions brought in by new members and leaders.

Though each of these mechanisms plays a crucial role, by far the most important for cultural beginnings is the impact of trust in those processes and values that the enterprise business has used so successfully so often that they give direction to its daily life.

To keep consistency with the leadership perspectives described in the previous chapter, we can consequently define an enterprise business culture simply as:

That character of genuine commitment and order in employees, teams and groups within the enterprise business that allows people to trust in those processes and values that the enterprise business has used so successfully so often that they give direction to its daily life and allow people to trust each other enough to work together.

In this definition, we can see that "trust" is a crucial aspect of an enterprise culture. We can think of trust as:

Expectancy held by an individual or group that promises will be kept and vulnerability will not be exploited.

It is an "expectation" of dependability and benign intentions typically viewed as a characteristic of personal relationships. Trust is a function of four distinct behavioral characteristics that together form the criteria for its assessment. These are:

1. Being honest (authenticity, forthrightness, veracity, sincerity)
2. Being dependable (reliability, consistency, follow-through)
3. Exercising judgment (ability, capability, capacity, decision making, wisdom), and
4. Generating partnership (Mutual Support, Shared Values and Concerns, Collaboration, Alliance Building).

Being able to fully trust another (whether or not that trust is ever verbalized) and trust in those processes and values that the enterprise business has used so successfully so often is a function of being genuinely satisfied with each of these four behavioral characteristics. To whatever degree trust is lacking, the source of the gap can always be traced to one or more of these four dimensions.

The development of trust involved being accountable for deviations from the expected performance. So long as those trusted behaved in line with expectations, trust would be reinforced as a result of experience and built progressively over time. It does not need to involve belief in the good character or morality of the individuals and groups involved, merely it needs their conformance to agreed action. It follows our acceptance of an assumed truth about another person or thing.

The development of trust continues and is sustained and enlarged only as future experiences confirm that early perception of expectations to be, in fact, correct. That is, trust builds as experience proves the essential truth of our initial perceptions. Trust diminishes by the reverse; as those trusted do not behaved in line with expectations, we withdraw our trust.

In enterprise businesses nowadays, trust is typically seen as outside the domain of most managers and even Human Resource departments. It is founded in core values and beliefs that have evolved over time in teams within the enterprise through the accumulation of actions and events the members of teams experienced. These core values are embedded in generally known and understood statements about what is good or not good in and about the enterprise.

The establishment of trust between people within an enterprise business has been shown to develop a sense of community and, for example, has encouraged people to work together with less control and obvious hierarchy (Misztal, 1996). In large-scale enterprise businesses, hierarchy could to some extent be replaced by trust and reputation. Jönsson studied the effects of trust on knowledge work. As the value of knowledge work increased and the work itself became more knowledge intensive, so the competence of command and control hierarchies to take decisions over that knowledge work became less feasible and less legitimate unless the decisions were taken in consultation with the relevant knowledge workers (Jönsson, 1996).

The development of trust involve being positively accountable for deviations from the expected performance. So long as those trusted behaved in line with expectations, trust would be reinforced as a result of experience and built progressively over time. It does not need to involve belief in the good character or morality of the other party, merely their conformance to agreed action. The replacement of a command and control approach with a more democratic and communicative approach has to be based on establishing trust between the individuals and groups involved.

From the perspective of "*character of genuine commitment and order in teams within the enterprise business that allows people to trust each other enough to work together*," we can see that leadership creates the culture within an enterprise business and management lives within that culture. The enterprise business culture, therefore, is about how much members trust each other, if indeed they trust others at all, to work together. It is about attitudes and emotions and their impact on the different teams performances.

4.3.1 Features of Enterprise Business Culture

Although an enterprise business culture is a potent force, it cannot fully dominate individuals' thought and action because of the capacity of human agents to comment critically on their situation and to choose to abstain or act otherwise than the enterprise business cultural norms would dictate. As such, an enterprise business culture explains how people perceive the enterprise business, and consequently determines how they behave.

In a large scale enterprise business, there is the potential for multiple and even competing subcultures existing within the enterprise business, each different in some respects from each other and from the parent culture. Therefore, the enterprise business culture could be viewed as the sum of many subcultures, each of which contributed its own nuances of meaning. Knowing the parameters of the larger culture, though, helps in defining and analyzing the details of the subcultures making up the larger body. Enterprise business culture, moreover, may often vary more across enterprises than within them, indicating that many cultural elements may not be unique to particular enterprise business in the same industry.

Although there is a widely held view that culture is the glue that could bind an enterprise business together (Deal & Kennedy, 1982), it is equally clear that culture could in fact be divisive, just as easily as cohesive. In the absence of any dominant super-culture, the various subcultures could well be in conflict with each other. On occasion this conflict might become overt and sometimes highly dysfunctional, but more usually the conflict would be bubbling below the surface.

An enterprise business culture could thus be described as a "mix of cross cutting subcultures," continually reacting against each other in some more or less cohesive, or divisive, not necessarily stable and equilibrium (Gregory, 1983). The suggestion that culture could in some way be controlled to make management's desired culture both coherent in itself and dominant over other subcultures is an essential quality of the excellent enterprises (Peters & Waterman, 2004).

In the management literature, there are many stories, even legends or myths about the great and good, or not so good, originators of such enterprises businesses as Toyota and GE family of companies. These enterprises businesses all share "strong," deliberately established and maintained, coherent, dominant "cultures." They have, it is argued, gained the active participation of all their members and their consistent concentration of effort on pushing their businesses further in its intended direction.

The distinctive feature of these strong cultures is that they are shared by all members within the enterprise business. The common assumptions about the enterprise business, and the way to behave in it, represented a powerful means of getting to the "hearts and minds" of all members. Thus, potentially, an enterprise culture offers a way for people in the enterprise business to focus their efforts, consistently to achieve a sense of direction and achievement, beyond the scope of more orthodox management approaches.

However, the creation of a strong culture is not without problems. Whilst it may serve to replace the control mechanisms effectively wielded by powerful bureaucracies and self-preserving functions with something that appeared more congenial, it could also replicate the problems created by those control mechanisms. Culture is by definition long lasting, and strong culture may be particularly so. Thus all the rigidities and loss of responsiveness that caused mechanistic enterprise businesses so much trouble when confronted with change, could be equally present in the strong culture enterprise businesses. Control is control, whether or not it came in the guise of a nineteenth-century bureaucratic structure, or a 1970s strong culture. In either format tight control implied lack of flexibility and ability to innovate.

Nevertheless, it is feasible that careful culture management could achieve what George Preston had referred to as the "true spirit of co-operation between employer and employee." Although he was speaking on behalf of labor, his plea surely applied through the organizational hierarchy of an enterprise business. Co-operation, rather than compliance or coercion, should surely always be the preferred way forward for people within an enterprise business, although from time to time cooperation would be bound to break down when interests are competing too directly.

Thus, in the process of moving the enterprise business as a whole (businesses and customers) from its current stage of maturity toward the "Continuous Improvement" stage of maturity, the prime task of the enterprise business executives, its managers and leaders, therefore, is to create a culture or sub-cultures that integrates all individuals into a natural unity of trust so that individual actions can strengthen the results of the whole. When the prevailing enterprise business culture or sub-culture is incompatible with the business vision, the task is to change it to ensure that it promotes needed integration and harmony.

The evidence of the last 25 years indicates that trying to develop successful enterprise cultures by copying or reproducing tools, techniques, or principles of a successful culture enterprise business does little to change an enterprise business' culture. For example, how do you get people to actually live principles? Tools and techniques, the things you see, are built upon invisible routines of trust, thinking and acting, particularly in management, that differ significantly from those found in most enterprise businesses.

On the other hand, focusing on developing daily behavior patterns on trust is a leverage point because, as the field of psychology shows us, with practice, behavior patterns are changeable, learnable, and reproducible. Until very recently, there has been little effort to measure trust as an organizational construct. Fortunately, Dean Spitzer has developed the questionnaire below, in Table 4.1, for measuring "trust" within enterprise businesses, based on extensive research (Spitzer, 2007).

Successful enterprise business cultures, characteristic of enterprises business at the "Continuous Improvement" stage of maturity, are characterized by enough mutual trust, respect and commitment to let employees be free to make choices, which empower them to meet at least some of their needs.

Command and control systems and structures typical of businesses at lower maturity stages seldom provide that trust or that freedom, except, perhaps, at the very top stages. Thus, building trust and respect, and maximizing commitment is an absolute must for any enterprise business to succeed in its renewal effort to attain the "Continuous Improvement" stage of maturity. Without it, valuable energy will be wasted.

The extent of the cultural transition is as broad and comprehensive as the enterprise itself. Some critical aspects of the enterprise life which have practical cultural implications include: communication as the nerve system of the enterprise; cooperation; conflict creation and resolution mechanisms; commitment; cohesiveness and member ownership of enterprise aims; levels of acceptable caring and concern for others; and ultimately, trust. These operating processes interact to form the social aspects of the enterprise.

Of course, any other aspect of the enterprise relationship or external factors impinging on teams performances can shape or modify the culture within an enterprise. Thus, employee's professionalism, personal or professional biases, and social or politics can be features of a culture within an enterprise. Similarly, task or system complexity, changing work values, training and development, task design,

Table 4.1 Enterprise business trust questionnaire

#	Observation	Rating
01	I trust the expectations that have been communicated in this enterprise/business unit/department/team	
02	I feel that people in this enterprise/business unit/department/team are honest	
03	There is mutual respect among members in this enterprise/business unit/department/team	
04	People in this enterprise/business unit/department/team are good at listening without making judgments	
05	I feel good about being a member of this enterprise/business unit/department/team	
06	I feel that the people in this enterprise/business unit/department/team are competent	
07	I feel confident that this enterprise/business unit/department/team has the ability to accomplish what it says it will do	
08	People help each other learn in this enterprise/business unit/department/team	
09	Learning is highly valued in this enterprise/business unit/department/team	
10	I feel that I can be completely honest in this enterprise/business unit/department/team	
11	Honesty is rewarded in this enterprise/business unit/department/team	
12	There are clear expectations and boundaries established in this enterprise/business unit/department/team	
13	Delegation is encouraged in this enterprise/business unit/department/team	
14	People keep agreements in this enterprise/business unit/department/team	
15	There is a strong sense of responsibility and accountability in this enterprise/business unit/department/team	
16	There is consistency between words and behavior in this enterprise/business unit/department/team	
17	There is open communication in this enterprise/business unit/department/team	
18	People tell the truth in this enterprise/business unit/department/team	
19	People are willing to admit mistakes in this enterprise/business unit/department/team	
20	People give and receive constructive feedback non-defensively in this enterprise/business unit/department/team	
21	People maintain confidentiality in this enterprise/business unit/department/team	
22	I can depend on people to do what they say in this enterprise/business unit/department/team	
23	People are treated fairly and justly in this enterprise/business unit/department/team	
24	People's opinions and feelings are taken seriously in this enterprise/business unit/department/team	
25	I feel confident that my trust will be reciprocated in this enterprise/business unit/department/team	
Use the standard five-point rating scale:		
5 = Strongly agree; 4 = Agree; 3 = Neither agree nor disagree; 2 = Disagree; 1 = Strongly disagree		
Interpretation key:		
Highest score is 125		
High score range is 100–125. Your enterprise business is doing a good job on its culture. Staff responds to stimulus because of their alignment to organizational values		
Moderate score range is 70–110. Your enterprise business culture is making progress towards a strong enterprise business culture		
Low score is below 70. Your enterprise business culture is still very much in need of improvement. There is little alignment with organizational values and control is exercised through extensive procedures and bureaucracy		
Danger zone is below 50		

and task assignments systems are also cultural determinants. Collectively, these cultural factors influence how people respond to the requirements of the work system employed and constitute the work culture.

Creating an enterprise culture involves leadership in several important mind-changing tasks. Among these tasks are activities for setting the value base for mutual interaction and thinking strategically about teams and their future within the enterprise. This perspective involves systematically shaping a desired culture within which members can trust others and expect others to trust them.

Strategic Planning and Management 5

The term "strategy" comes from the Greek word "strategos," deriving etymologically from stratos (the army) and agein (to lead). Thus, in this original sense, "strategy" is "the art of leading the army." While it has originated in the military sphere, the term "strategy" has risen into prominence in the business world since the 1960s to become a cornerstone of high-performing enterprises nowadays.

This chapter delves into the key characteristics and constituents of an enterprise business strategic planning and management necessary to take your enterprise business to the "Continuous Improvement" stage of maturity as described in a previous chapter.

5.1 Enterprise Business Intended Strategy

In business practices, an enterprise business intended strategy is an integrated concept represented by long-term guidelines on the enterprise business as a whole or at important parts of its business, with the intended objective of ensuring survival through target market positions and competitive advantages that the enterprise business must build and maintain. It is determined by the enterprise business executives and management, and it describes the potentials success which the enterprise business must be built up or maintained. An enterprise business intended strategy also evolves on a dynamic basis over time in response to the internal and external contingencies that emerge to confront the enterprise business.

Mintzberg defines an enterprise business intended strategy in terms of 5Ps (Mintzberg, 1978). These 5Ps are:

1. *Perspective* – It is the basic business concept or idea, and the way in which that concept or idea is put into practice (or implemented).
2. *Plan* – It is a direction, a guide, or a course of action from now (or from the past) into the future, however that "future" is defined and whatever the time horizons associated with it.
3. *Pattern* – It is the consistency of enterprise decision-making over time.

A. Van Aartsengel and S. Kurtoglu, *A Guide to Continuous Improvement Transformation*, Management for Professionals, DOI 10.1007/978-3-642-35904-0_5,

4. *Position or positioning* – It is an indication by which the enterprise business "locates" itself within its external and competitive environments; and by which it positions specific products or services (and therefore the resources and capabilities required to produce them) against the demands of the market segments it serves.
5. *Ploys* – These are the competitive moves or competition strategies designed to maintain, reinforce, achieve, or improve the relative competitive position of the enterprise business within its sector and markets.

Enterprise business intended strategies are the means by which the enterprise business intends to achieve its objectives. They describe the chosen "paths to goal," or "routes to achievement" or "plans of campaign." They act as "ground-rules," and define the nature and occasion of the decisions needed to achieve enterprise objectives. Intended strategies have their time scale and their risk content, and they determine how the enterprise intends to carry out its activities during the time horizons to which it is working, in order to achieve its objectives.

As economic entities, enterprise businesses need intended strategies in order to: set their priorities as regards to resource allocation; be able to react to uncertainty and turbulence: changes in their environment; respond to competitors' behavior; or communicate the direction of their own business to employees, customers, and shareholders.

There are many forces contributing to uncertainty and turbulence in today's economy. Some are macroeconomic, other forces are political and regulatory, and other forces are more industry or even company-specific. In some industries, rapid consolidation is forcing companies to consider multiple scenarios and choose what role they want to play in the market place. In other industries, the rise of rapidly developing economies, such as China, India, and Brazil, is creating new competitors and more complex competitive dynamics. And even in more stable industries, many enterprise face uncertainty and turbulence simply owing to the fact that their existing portfolio of businesses is maturing, requiring the development of new business models and new platforms for growth.

Not every enterprise business has to struggle with all these forces, which contribute to uncertainty and turbulence. But most will face at least some of them; and relatively few will encounter none at all. It is important to keep in mind, however, that while turbulence presents majors challenges, it also creates opportunities. For that reason, effective intended strategy will be the key to superior value creation in the long term.

Because of the particular intended strategies they adopt at the global, corporate, business, and functional levels, some enterprise businesses in very tough industries (automobile and turbo-machinery for example, where we have been fortunate to provide consulting services) consistently deliver higher performance than their competitors. As indicated in a previous chapter, these successful enterprises balance two needs:

1. The need to look backward in order to maintain the existing enterprise business and its current customers, and
2. The need to look forward in order to explore and achieve performance breakthroughs and to identify and attract new customers and new sources of value. This second need is the purpose of strategic planning.

Emerging strategic imperatives include goals of not just aiming to optimize within the enterprise business' current industry but trying to generate ground-breaking intended strategies which will create new niches and markets and re-define whole industries. In other words, ensuring survival through target market positions and competitive advantages not just for market share in existing markets but for "opportunity share" in future markets, in a world of continuous re-definition of industry boundaries and commingling technologies. This implies that enterprise businesses should not merely or exclusively be trying to catch up with the best performers, as Toyota's and GE's family of companies, in current competitive markets, but be aiming to invent new markets, re-write the rules and create new competitive space. The development of these intended strategies is a very complex task that requires competence in a number of different functional areas within the enterprise business.

5.2 Strategic Management

Strategic management is a set of activities followed to establish and implement an enterprise business intended strategy. The establishment of intended strategies, the making of plans, and the implementation of those intended strategies and plans are key management decision-making processes in any kind of enterprise business. The strategic decision-making process takes place, in some form or other, in most kinds of businesses. The process may be formalized and systematic. Or it may be informal, opportunistic and ad hoc in nature.

Strategic management is concerned with management planning and decision-making for the medium to long-term future. It is concerned with the anticipation of that future, and with the establishment of a vision or view of how the enterprise should develop into the future that it must face. Strategic management is also concerned with the character and direction of the enterprise business as a whole. It is concerned with basic decisions about what the enterprise business is now, and what it is to be in the future. It determines the purpose of the enterprise business and provides the framework for decisions about people, leadership, customers or clients, risk, finance, resources, products, systems, technologies, location, competition, and time. It determines what the enterprise business should be capable of achieving, and what it will not choose to do. It will determine whether and how the organization will add value, and what form that added value should take.

The tasks involved in strategic management can be broken down into strategic planning, the implementation of intended strategies, and strategic control. Strategic planning, which is the process by which strategies are produced, forms the basis for the implementation of intended strategies and strategic control. A systematic approach to strategic planning, which is firmly grounded on realities, is seen by many enterprise leaders and management researchers as an essential requirement for long-term enterprise success (Aaker, 1992; Abraham, 2012; Barksdale & Lund, 2006; Bradford, Duncan, & Tarcy, 2000; Chambers & Taylor, 1999; Espy, 1986; Fogg, 1994; Goodstein, Nolan, & Pfeiffer, 1993; Grünig, Kühn, & Clark, 2010; Kaufman, Oakley-Browne, & Watkins, 2003; Mintzberg, 1994; Rea & Kerzner, 1997; Simerson, 2011; Wittmann & Reuter, 2008; Wootton & Horne, 1997).

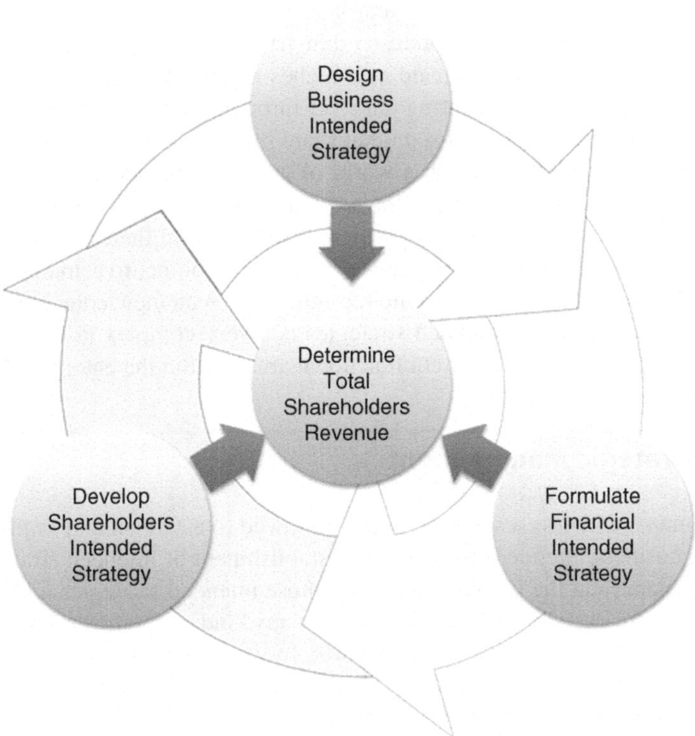

Fig. 5.1 Development process of an enterprise intended strategy

5.3 Strategic Planning Process

Ideally, an enterprise business intended strategy planning defines the fundamental logic that explains why a particular set of businesses are set together in the first place within the enterprise. For example, it should identify the parenting advantage and operational synergies that make the enterprise the best owner of its particular set of businesses. And it should define the precise role of each of its businesses in the enterprise business' overall value-creation intended strategy.

Enterprise business intended strategy planning is also responsible for making sure that the enterprise business' portfolio of businesses evolves over time. Some businesses inevitably mature and may no longer be able to create value at a level that matches the enterprise business' aspirations. Thus, the enterprise business intended strategy planning must also consider the full range of factors affecting the total shareholders returns (TSR).

An enterprise business intended strategy and the value that the enterprise business creates exist in a symbiotic relationship, as illustrated in Fig. 5.1. An enterprise business intended strategy defines the key areas of the enterprise business competitive advantages and how it will exploit those advantages to create value to its shareholders. But the energy flows in the opposite direction as well. Value creation is an important foundation for future competitive advantage. Not only does it satisfy shareholders demands, it also solidifies their support for management's long-term planning.

Figure 5.1 depicts a broader approach to developing an enterprise intended strategy. The process places value creation at the center of the development of the enterprise intended strategy. It supplements the focus on the sub-process for the design of business intended strategy with sub-processes for strategy focusing on the enterprise financial policies and shareholders priorities and goals.

This approach originates from the "Boston Consulting Group" in its development series on "value creators" (Olsen, Plaschke, & Stelter, 2008, 2009; Olsen, Plaschke, Stelter, & Farag, 2011). The advantages of this approach to developing an enterprise business intended strategy include:

1. *Simultaneous and tandem planning* – planning and decision making flow in tandem and simultaneously. Once the intended business strategy (including that of its various component businesses) is defined and specific financial targets are set, those choices then determine the parameters of the enterprise financial policies and the communications necessary to establish the intended strategy to the shareholders.
2. *Non sequential (hence self-interest) decision making* – Because the decision making is non sequential, it reduces self-interest fragmentation across different operational and functional businesses. In this approach, corporate strategists, corporate finance, and shareholders relations work together in tandem, within the project team designed to develop the enterprise intended strategy, to produce an objective fact-based analysis of what it will take to create value from the enterprise business.
3. *Strong connection to value creation* – Few enterprise businesses have an explicit goal for shareholders value. And those that do rarely incorporate that goal explicitly in their planning process or quantify the potential total shareholder returns contribution of their business plans. As a result, value creation may be a desired outcome, but it is not an actual driver of the intended strategy development.

Using this approach, business intended strategy, financial intended strategy, and shareholder intended strategy are examined in tandem and simultaneously (not sequentially) by the entire enterprise project team designed to develop the enterprise business intended strategy (not by isolated functional experts) in order to identify and reach agreement on critical tradeoffs.

The proposed process for developing an enterprise business intended strategy takes into account these interactions and linkages and allows enterprise business executives to manage the tradeoffs among them, focusing on value creation.

A key aspect of this approach is to view business intended strategy, financial intended strategy, and shareholders intended strategy as three equal parts of an enterprise intended strategy and to thread them in tandem and simultaneously rather than sequentially. This integrated perspective is critical because both enterprise financial policies and the goals and priorities of its dominant shareholders can have important implications for the each of the enterprise businesses intended strategies

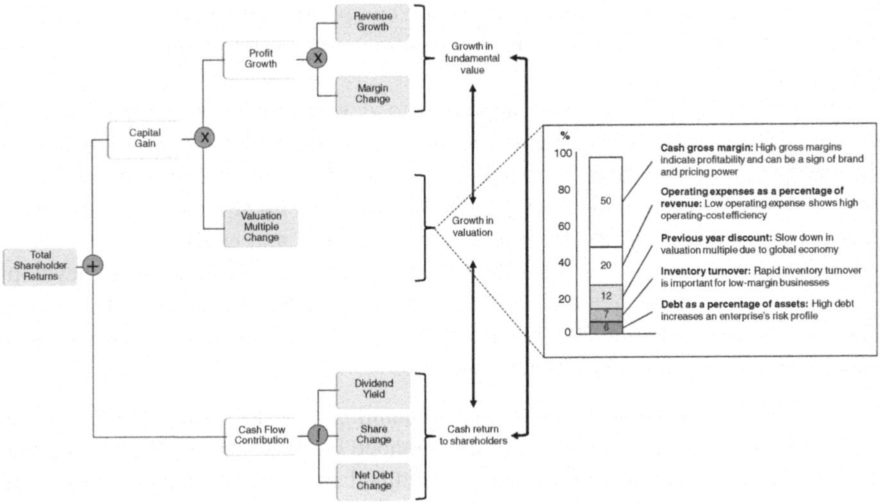

Fig. 5.2 Drivers of total shareholder returns, as indicated by The Boston Consulting Group

(and vice versa). They also can have a direct – and, sometimes, quite substantial – impact on the total shareholder returns in their own right.

Most enterprise business executive teams believe that they are already committed to increasing shareholders returns. After all, they talk about it all the time during lively presentation and communication to employees. Some may even have set a target for improvement in total shareholder returns. But in most cases, they are not really focused on value creation for these shareholders or for any value creation at all, because their enterprise business intended strategy planning process does not consider the full range of factors affecting the total shareholder returns.

In its "Value Creators" reports series (Olsen et al., 2008, 2009; Olsen, Plaschke, & Stelter, 2010; Olsen et al., 2011), the Boston Consulting Group has introduced an integrated model for determining the total shareholder returns. The model, illustrated in Fig. 5.2, incorporates three critical dimensions:

1. The first is improvement in fundamental value, represented by the discounted value of the future cash flows of an enterprise business based on its margins, asset productivity, growth, and cost of capital.
2. The second is improvements in an enterprise business valuation multiple, driven by shareholders expectations that shape how capital markets value an enterprise fundamental performance at any given moment in time.
3. The third is the direct payment to shareholders or debt holders in the form of dividends, share repurchases, or the pay-down of debt.

The key point about this model for determining the total shareholder returns is that these three dimensions exist in dynamic interaction. For example, an enterprise business may improve its fundamental value through profitable growth. But precisely how the enterprise business goes about achieving that growth can have either a positive or a negative impact on its valuation multiple and, therefore on its total shareholder returns.

Alternatively, the level of the enterprise business multiple, compared with those of its peers, can enable certain business strategies and make others impossible. For instance, an especially strong multiple can make the enterprise stock a handy currency for acquisition; conversely, a weak multiple can make the enterprise vulnerable to take-over.

Finally, cash payouts not only can contribute directly to the total shareholder returns but also can have a positive impact on the enterprise multiple by both strengthening the loyalty of existing shareholders and attracting new investors.

5.3.1 Financial and Shareholders Intended Strategies

The financial intended strategy is the result of many different decisions about issues such as the enterprise capital structure, preferred credit rating, dividend policy, share repurchase plan, tax intended strategy, and hurdle rates for investment projects or mergers and acquisitions (M&A). Often these seem like discrete issues. But it takes a holistic approach to optimize the overall financial intended strategy.

For example, consider the impact of an enterprise business unit's proposed growth initiative. Business unit managers will naturally be focused on the initiative's return on investment – that is, whether it has a positive net present value (NPV). But even when a proposed growth initiative delivers returns above the cost of capital, the enterprise may have been able to get even greater returns by, for instance, returning the cash to shareholders.

Enterprise businesses that are overleveraged, that are undervalued compared with their future plans, or that suffer from a low valuation multiple relative to peers can often realize major improvements in their valuation multiples and total shareholder returns by paying out more cash to shareholders or by using that cash to reduce debt. In simple terms, every investment option needs to be considered simultaneously against alternative uses of capital. Unless the enterprise business project team designed to develop the enterprise business intended strategy (not by isolated functional experts) integrate considerations of business intended strategy, managing such tradeoffs is extremely difficult.

Similarly, with the enterprise shareholders demands and priorities, it is essential for the enterprise business intended strategy to be aligned with priorities and expectations of its shareholders. Those expectations will drive the enterprise business valuation multiple relative to its peers, which is the key source of short-term total shareholder returns and a critical influence on the enterprise long-term value creation.

5.3.2 Design Business Intended Strategy

A process-based approach for the design of an enterprise business intended strategy, which we highly recommend to enterprise business executives and managers engage in strategic planning implementation, has been developed by Grünig and Kühn (Grünig et al., 2010). It offers a comprehensive system of strategic thinking with uniform terminology and combines the most important methodological approaches within a single recommended planning process.

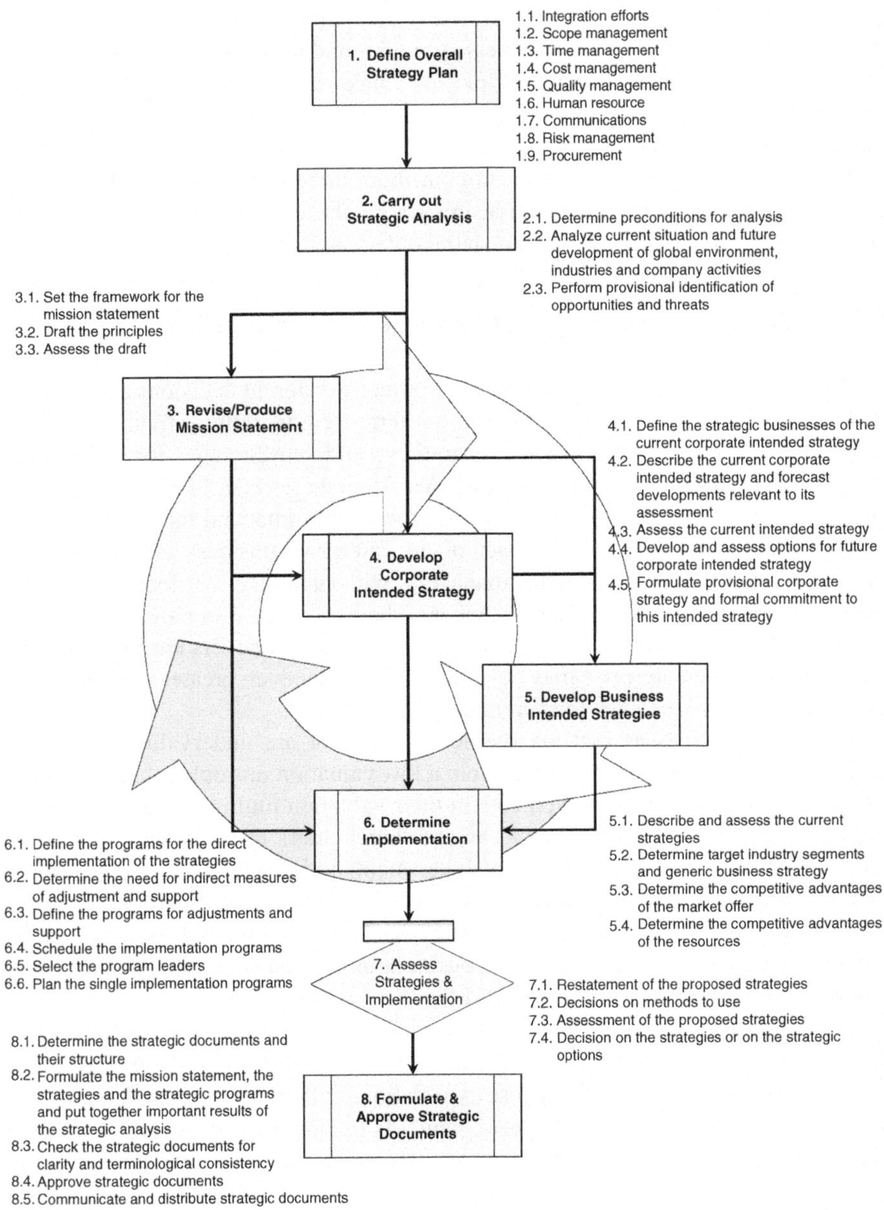

Fig. 5.3 Generic strategic planning process

Within a project management context, the developed strategic planning process, illustrated in Fig. 5.3, includes the actions necessary to define, analyze, develop, implement, assess and formulate all subsidiary plans into strategic documents. This section condenses and describes the steps of the strategic planning process.

5.3.2.1 Define Overall Strategy Plan

The initial step of the strategic planning process consists of activities that integrate the various elements within the strategic planning process. This integration includes characteristics of unification, consolidation, articulation, and integrative actions that are crucial to completion of the strategic planning, successfully meeting customers and stakeholders requirements, and managing expectations. Within this step, choices are made about where to concentrate resources and effort on any given day, anticipating potential issues, dealing with these issues before they become critical, and coordinating work for the overall strategic planning good.

The main objective of this initial step is to effectively integrate the activities that are required to accomplish the strategic planning objectives within the enterprise business defined procedures. These activities include:

1. Developing the strategic planning project charter that formally authorize strategic planning project.
2. Developing the preliminary scope statement that provides a high-level scope narrative.
3. Documenting the actions necessary to define, prepare, integrate, and coordinate all subsidiary plans into a strategic planning project management plan.
4. Executing the work defined in the strategic planning project management plan to achieve the requirements defined in the scope statement.
5. Monitoring and controlling the activities used to initiate, plan, execute, and close the strategic planning project or a phase of it to meet the performance objectives defined in the plan.
6. Reviewing all alteration requests, approving alterations, and controlling alterations to the deliverables.
7. Finalizing all activities to formally close the strategic planning project or a phase of it.

5.3.2.2 Carry Out Strategy Analysis

The second step of the strategic planning process aims to provide a provisional picture of the current situation of the enterprise business and the possible developments. It is concerned with:

1. Determining the preconditions for analysis:
   - Determining the markets and activities to be analyzed
   - Determining the methods and the resulting data quality
2. Analyzing the current situation and the future development of global environment, industries and company activities:
   - Analyzing global environment with global environment analysis
   - Analyzing the industry markets with market system analysis and identification of strategic success factors
   - Analyzing the company activities with strengths and weaknesses analysis and, if necessary, stakeholder value analysis

3. Provisional identification of opportunities and threats:
   – Provisional identification of the opportunities and threats at corporate level
   – Provisional identification of the opportunities and threats for each activity

In this step, the enterprise business examines its own internal or corporate characteristics and capabilities; and identifies the most important features of the external environment within which it must operate. This second step emphasizes on data collection, structuring, and analysis in three fields: global environment, specific task environments or industries, and the enterprise itself. The data is collected only to the extent that it is needed to answer strategic questions

Grünig and Kühn consider the environment under five categories: economic conditions and developments, social and cultural developments, ecological developments, technological change, and political and legal developments. Documenting changes in these categories must involve analysis of the regulatory bodies, which include: state institutions, employers' and industry groupings, trade unions and consumer organizations.

Data collected from these regulatory bodies must be integrated into the five categories for the environmental analysis. At this step of the process, five concrete methods, which culminate in the identification of threats and opportunities, both at the level of individual activities and for the enterprise as a whole, are recommended for use in the analysis:

1. Global environmental analysis, which focuses on developments in the company's environment, identifying trends which could have a major impact on the company's situation.
2. Market system analysis or analysis of the value creation chain of the industry.
3. Identification of strategic success factors or identification of criteria for customer choice. These are variables which have important effects on long-term success. As well as a number of general success factors which apply in all industries, each industry has its own industry-specific success factors. The success factors are important because they reveal the dimensions of competition. It is in these important dimensions that competitive advantages can be constructed.
4. Strengths and weaknesses analysis or competitor analysis or benchmarking.
5. Stakeholder value analysis

5.3.2.3 Revise or Produce Mission Statement

The third step of the strategic planning process is concerned with the revision or production of the enterprise business mission statement. It involves setting the framework for the mission statement, drafting the principles, and assessing the draft.

5.3.2.4 Develop Corporate Intended Strategy

The fourth step of the strategic planning process is the development of the corporate intended strategy (what industries/markets should the enterprise operate in?), which must guarantee that the enterprise business will direct its activities at attractive markets where it can build or maintain an advantageous competitive position. This step involves: defining the current strategic businesses (in what segments should the enterprise compete – and how?), a strategic business being a three dimensional construct which identifies a particular market, specific market offers and specific resources; describing and assessing the current market positions of the businesses; and determining target market positions and investment priorities for the businesses.

In large enterprise businesses, it may be convenient to divide strategy formulation into two inter-related components: the corporate strategy (what industries/ markets should the enterprise business operate in?) and the business strategies (in what segments should the enterprise business compete – and how?). The corporate strategy deals with issues of strategic management at the level of the enterprise business as a whole. Such issues will include the basic character, capability, and competence of the enterprise business; the direction in which it should develop its activity; the nature of its internal architecture, governance and structure; and the nature of its relationships with its sector, its competitors, and the wider environment. The business strategies deal which the enterprise strategies for specific business or organizational activities, specific sectors and markets, and specific divisions or business units into which operations are allocated.

Strategy development in the small to medium sized enterprise is unlikely to differentiate corporate and business strategies. The corporate strategy of the enterprise business is its business strategy, at least until the small to medium sized enterprise business grows to a sufficient size to have to think about issues of corporate development and external relationships.

5.3.2.5 Develop Business Intended Strategy

The fifth step of the strategic planning process is concerned with developing the business strategies. These business strategies specify the resources and offers which are needed for each business so that it can achieve or protect the target market positions set out in the corporate strategy. To this end, the business strategies identify the competitive advantages which have to be built up or maintained. Furthermore, to function as a long-term framework for the development of a business, a business strategy has to answer the following questions:

1. What customer groups will be served and what types of products and services should be offered to them?
2. Which generic business strategy will be followed to do this?
3. What competitive advantages will have to be built up on the level of the market offer?
4. What resources will be required to maintain or upgrade these competitive advantages?

The activities in this fifth step include: describing and assessing the current strategies; determining target industry segments and generic business strategy; determining the competitive advantages of the market offer; and determining the competitive advantages of the resources.

5.3.2.6 Determine Implementation

The sixth step of the strategic planning process is concerned with the planning of the implementation measures, especially the creation of the strategic programs. It includes activities related to defining the implementation programs, planning the implementation programs, and setting budgets the implementation programs.

5.3.2.7 Assess Strategies and Implementation

The seventh step of the strategic planning process is concerned is the global assessment of both strategies and programs. In this step, the overall value of the strategies, the fit with the overriding objectives and values, and the feasibility of the programs are assessed in terms of resources, risk and cost involved.

5.3.2.8 Formulate and Approve Strategic Documents

The eighth and final step of the strategic planning process is concerned the formulation and approval of strategic documents, which also form the basis of strategic control. This includes activities related to:

1. Determining the strategic documents and their structure:
   - Determining the strategic documents
   - Determining the structure of each strategic document
2. Formulating the mission statement, the strategies and the strategic programs and putting together important results of the strategic analysis
3. Checking the strategic documents for clarity and terminological consistency:
   - Checking for terminological consistency
   - Checking for clarity
4. Approval of the strategic documents
5. Communication and distribution of the strategic documents:
   - Communication
   - Distribution

In business practice, strategic aims and measures are normally set out in a variety of types of strategic plan. There can be considerable differences, both in the names given to these plans and in what they typically contain. The most important strategic documents are identified as: corporate strategies and business strategies. Other important documents are mission statements and functional area strategies.

5.4 Strategic Control

As we have seen, strategic planning sets out long-term goals and provides an intended guide to what is necessary in terms of actions and resources. This provides a clear direction and basis for the strategic implementation. Strategy implementation deals with the realization of intended strategies at the material level of market offers and hard resources, but it also includes complementary measures concerning personnel. Unsatisfactory implementation at the personnel level is the most frequent cause of failure in achieving success with intended strategies.

Strategic control, which closes the tasks involved in strategic management, has a dual function: to provide feedback on how strategies are realized and to check whether the assumptions or premises underlying the strategic plans correspond to reality. If there is too much divergence between the intended strategic plans and their implementation, or if the premises behind the intended strategies do not correspond to reality, then planning must begin again.

There are three elements in the strategic control: strategic realization checking, strategic monitoring and strategic scanning. Realization checking serves to guarantee that intended strategic measures are realized. Strategic monitoring begins after formulation of the intended strategy. Strategic scanning involves the global intuitive observation of the environment.

5.5 Conclusion

As mentioned already, enterprise businesses as economic entities need intended strategies in order to: set their priorities as regards to resource allocation; be able to react to changes in their environment; respond to competitors' behavior; or communicate the direction of their own business to employees, customers, and shareholders. In practice, however, it is hardly possible to realize intended strategies completely due to changing environment and business landscapes.

Thus, the realized strategies normally diverge to a greater or lesser extent from the intended strategies. Enterprise businesses at the "Continuous Improvement" stage of maturity periodically assess and measure the extent to which the realized strategies deviate from the intended strategies and incrementally close the gap between the intended and realized strategies. Furthermore, in these enterprise businesses, strategy drives the pattern of decisions within the enterprise business as a whole.

6 Performance Measurement

Within the context of this book, the word "performance" refers to how well a person, a group of individuals, a machine, a system, etc. does a piece of work or an activity. In the previous chapter, we have illustrated that an enterprise intended strategy determines the intended purpose of the enterprise and provides the framework for decisions about people, leadership, customers or clients, risk, finance, resources, products, systems, technologies, location, competition, and time.

This chapter delves into the key characteristics and constituents of performance measurement necessary to take your enterprise business to the "Continuous Improvement" stage of maturity as described in a previous chapter.

In practice, an enterprise business intended strategy is realized to a lesser or greater extent. How well an enterprise business realizes its intended strategy defines the enterprise business performance. Implicit in this definition are a criterion of success and an actual measure of success.

6.1 Performance Measure

We define an enterprise business "performance measure" as "*a criterion of success stated in relation to the enterprise business realized strategy or in relation to its intended strategy.*"

Thus, the goal of a "performance measure" is to enable improvement. For example, a multi-national enterprise has this commitment in its intended strategy: "*Prepare the assembly business unit for the transition to a continuous improvement maturity stage by ensuring the executives at all levels understand their new responsibilities and accountabilities,*" the performance measures are:

1. Training is provided to executive managers across the assembly business unit.
2. A communications plan is created for the assembly business unit and is used to educate managers and employees on continuous improvement maturity
3. Assembly business unit executives are working to integrate HR planning into business planning

A. Van Aartsengel and S. Kurtoglu, *A Guide to Continuous Improvement Transformation*, Management for Professionals, DOI 10.1007/978-3-642-35904-0_6,

Dean Spitzer, in his breakthrough book "Transforming Performance Measurement" (Spitzer, 2007), written about performance measurement, presents an essential approach to performance measures for enterprise businesses to achieve effective transformational performance. He describes how performance measures, properly understood and implemented, can transform organizational performance by empowering and motivating individuals. Performance measures understood in this way moves beyond a traditional view of quick fixes and fads to sustainable processes that can be used successfully to coordinate decisions and actions uniformly throughout an enterprise business.

Dean R. Spitzer shows that – and we concur with this view – an enterprise performance measure is not primarily about the numbers, but about providing clearer perception and greater shared insight and knowledge within the enterprise business. Accordingly, an enterprise business "performance measure" should not be confused with an enterprise business "performance indicator," which we define as:

> *An actual specific measure (quantity or quality) of success stated in relation to the enterprise realized strategy or in relation to its intended strategy.*

For example, "improvement in customer satisfaction" is a standard performance measure for quality programs within enterprise businesses. The actual improvements that an enterprise business can make in satisfying its customers can be quantified with an indicator such as the "customer satisfaction index score."

As another example, "improvement in physical health" is a standard performance measure for exercise programs within training institutes. The actual improvements that customers of such institutes can make in their physical health can be quantified with indicators such as "blood pressure," "heart rate" and "stress test scores."

6.1.1 Major Functions of Performance Measures

Employees perceive, understand, and develop insight on the performance of the enterprise business through the enterprise business "performance measures." They are critical components for moving an enterprise business from its current maturity stage to a higher maturity stage, hence improving the enterprise business performance. Some of the major functions of "performance measures" within an enterprise are to (Spitzer, 2007):

1. Increase visibility of the enterprise business performance
2. Focus attention
3. Clarify expectations and get right to the point
4. Enable positive accountability
5. Improve execution
6. Promote consistency
7. Facilitate feedback
8. Enable strategic alignment
9. Improve decision making
10. Enable change and individual improvement

Increase visibility of the enterprise performance – Because most of what happens within an enterprise business (activities, processes, capabilities, and performance) is not directly visible, a performance measure then becomes "the enterprise eyes" so that activities, processes, capabilities, and performance can be effectively managed and improved.

Focus attention – Because employees are faced with so many competing demands on their time and resources, activities on which "performance measures" have been defined tend to get their attention.

Clarify expectations and get right to the point – One of the most important roles of enterprise business executives and managers is to communicate expectations to the workforce. Too often these are vague, resulting in considerable confusion.

Enable positive accountability – By positive accountability we mean an opportunity to perform and improve. An enterprise business "performance measure" will tell the employees how well they are performing against commitments. When employees can clearly see a performance measure in a way that is meaningful to them, they are much more likely to take positive action.

Improve execution – An enterprise business will not be able to realize its intended strategy without good execution. And it will not be able to execute well, consistently, without "performance measures." To synthesize Larry Bossidy and Ram Charan views, execution is a specific set of behaviors and techniques that companies need to master in order to have competitive advantage (Bossidy, Charan, & Burck 2002).

You cannot execute well without robust dialogue. Robust dialogue starts when people go in with open minds. They are not trapped by preconceptions or armed with a private agenda. They want to hear new information and choose the best alternatives, so they listen to all sides of the debate and make their own contributions. When people speak candidly, they express their real opinions, not those that will please the power players or maintain harmony.

Formality suppresses dialogue; informality encourages it. Formal conversations and presentations leave little room for debate. They suggest that everything is scripted and predetermined. Informal dialogue is open. It invites questions, encouraging spontaneity and critical thinking.

Informality gets the truth out. Finally, robust dialogue ends with closure. At the end of the meeting, people agree about what each person has to do and when. They have committed to it in an open forum; they are accountable for the outcomes.

Think about the meetings that you, as enterprise executive, manager, leader, or team member have attended – those that were a hopeless waste of time and those that produced energy and great results. What was the difference? . . . the difference was in the quality of the dialogue.

Robust dialogue alters the psychology of a group. It can either expand a group's capacity in executing tasks or shrink it. It can be energizing or energy-draining. It can create self-confidence and optimism, or it can produce pessimism. It can create unity, or it can create bitter factions. Robust dialogue brings out reality, even when that reality makes people uncomfortable, because it has purpose and meaning.

Follow-through is the cornerstone of execution, and every individual who is good at executing follows through religiously. Never finish a meeting without clarifying what the follow-through will be, who will do it, when and how they will do it, what resources they will use, and how and when the next review will take place and with whom.

And never launch a performance measurement and management initiative unless you are personally committed to it and prepared to see it through until it is embedded in the culture of the enterprise business.

Promote consistency – Outstanding enterprise business performance is not about success in a quarter or year; it is about consistent success over the long-term – and this requires more than good luck.

Facilitate feedback – An enterprise business will not be able to perceive early warning signals, diagnose early problems, execute anything or realize its intended strategy consistently without good feedback. Feedback is the basic navigational or steering device of any individual or enterprise. Without a good "performance measure" the enterprise is flying blind, and it will take a lot of good luck to realize its intended strategy or continually close the gap between its realized strategy and its intended strategy. In this perspective and as Dean Spitzer conveys, we can think of an enterprise business performance measure as the lock, and feedback as the key to improvement. Without their interaction, the door to improvement will remain closed.

Enable strategic alignment – Consistent behavior and performance across any enterprise business is impossible without an alignment to its intended strategy. In fact, most enterprise businesses nowadays appear to be composed of a collection of functional silos that operate so independently that there seems to be little connection among them at all, although employees are trying to do perform their tasks at best by maximizing their own functional measures of success. The key to making self-interest coincide with enterprise interest is through a full-alignment with the enterprise business intended strategy. We will discuss how in a next chapter of this book.

Improve decision making – Taking good decisions and taking them faster and more consistently than the competition is a characteristic of highly effective enterprises.

Enable change and individual improvement – When individuals want to change, they commonly use performance measures to help them to do so. In fact, when individuals feel good about their performance potential, they tend to want as much information as possible about how they are performing. They realize that performance measures are the key to improvement of their status. In a relatively nonthreatening environment, "performance measures" help employees track their progress toward a defined commitment. Project team members work energetically to reach predefined milestone, while open-ended timeframes lead them inevitably to complacency and low energy.

6.2 Realizing Performance Measurement

Within an enterprise business, each one of these functions of "performance measures" can be experienced in a positive or a negative manner depending on the maturity stage of the enterprise business. For example, each and every function will be experienced quite differently in within an enterprise business at the first or at the second stage of maturity, where leaders dictate or a command-and-control environment prevails. If employees perceive that a performance measure is in place to help them to become more successful (rather than to monitor, judge, command and control them) and to empower them (rather than coerce, manipulate and control them), then the performance measure will become a powerfully positive force in the enterprise.

In the process of moving the enterprise as a whole (businesses and customers) from its current stage of maturity toward the "Continuous Improvement" stage of maturity, a critical the task of every employee assuming a leadership role (as defined in a previous chapter), is also to create an optimal environment for effective use of the enterprise business performance measures.

As Dean Spitzer indicates (Spitzer, 2007), also confirm through our experience with client organizations, most enterprise businesses are unable to establish the right environment for the effective use of performance measures. This aspect is very crucial to developing and implementing performance measures within an enterprise business. In the following sub-sections, we describe and synthesize, following the writings of Dean Spitzer, the keys necessary to establish the right environment and take advantage of at least some of the functionality that "performance measures" have to offer within an enterprise business.

We will discuss the technical aspects of performance indicators (collecting data, calculations, analyzing data, statistics, etc.) in our next book entitled "Handbook on Continuous Improvement Transformation: The Lean Six Sigma Framework and Systematic Methodology for Implementation."

The four determining factors to making progress on the development and implementation of performance measures, as articulated by Dean Spitzer and following extensive analysis of existing leading organizations are: Context, Focus, Integration, and Interactivity. The following sections synthesize Dean Spitzer's writings.

6.2.1 Create a Positive Context for "Performance Measures"

Context, here, refers to the circumstances that form the setting for events, statements, ideas, constraints, data, social climate, or human factors, and in terms of which it can be fully understood and assessed, and within which "Performance Measurement" are carried out. The context of a performance measure, as with any improvement initiative, sets the tone by presenting the purpose of a performance measure as either negative (when used to inspect, control, report, coerce, manipulate) or positive (when used to give feedback, learn, improve). It reflects how the performance measure is perceived by employees and therefore how they respond emotionally to it. Furthermore, the context of a performance measure can make the difference between employees being energized by a performance measure or employees just minimally complying with it, and even using a performance measure for their own personal benefit. This is why the first determining factor in the development and implementation of performance measures is to create a positive context.

We have indicated in the section above that an enterprise business performance measure is not primarily about the numbers or the performance indicators, but about providing clearer perception and greater shared insight and knowledge within the enterprise business. A performance measure is implemented in a social, economic and environmental context within the enterprise, and has intended or unintended positive and/or negative impacts. As such, enterprise business executives, managers and leaders should consider a performance measure in its cultural and social environmental context, which largely determines its effectiveness. Several factors can affect this context, three of which strongly influence it. These are: the climate within the enterprise business, the measurement expectations, and the human factor.

The social climate within the enterprise business – It is the prevailing "atmosphere" within the enterprise business, the social-psychological environment that profoundly influences all behavior, and it is typically measured by employees' perceptions. The prevailing "atmosphere" is what best "defines" an enterprise business to employees. It reflects perceptions on a variety of dimensions, including, among others:

1. The extent of formality (hierarchical structure) versus informality
2. Trust versus distrust (and cynicism) of employees
3. Open versus closed communication
4. Controlling versus collaborative decision making
5. Inward-looking versus outward-looking
6. Past focus versus future focus
7. Task-focus versus people-focus
8. Change versus rigidity
9. Risk-taking versus risk aversion

Enterprise businesses at the "Continuous Improvement" stage of maturity are characterized by a prevailing "atmosphere" that is most conducive to the continuous development and improvement of performance measures. These enterprise businesses tend to be rated highly in such dimensions as openness, trust, honesty, collaboration, customer-focus, and flexibility.

The measurement expectations – Dean Spitzer propounded the view that the measurement expectations describe the performance measurement practices, and the "rules" of conduct relative to performance measures within the enterprise. Although not always explicitly documented, and often unwritten, these expectations tend to reflect the enterprise's assumptions, its deeply-held beliefs about performance measures. For example, expectations will prescribe what types of performance measures are most credible. In most organizations today, financial measures are still much more highly valued than nonfinancial ones. In some organizations, nonfinancial measurement is still just an afterthought, or even actively resisted.

The Human Factor – This is the most critical component of the context of measurement. The ideas and inspirations that guide and improve performance measures within the enterprise businesses come from its people. Indicators of performance measures; i.e. the actual specific measures, are of no value without human involvement. It is people who will ultimately determine their effectiveness. It is human beings, not machines, who transform performance indicators into information, information into insight, insight into knowledge, and knowledge into wisdom. Without people, indicators of performance measures would just sit in a repository or in a report and be good for nothing. People bring attitudes, commitment, capabilities, skills, knowledge, and experience into their use of the performance measurement.

Within enterprise businesses at lower stage of maturity, initiatives on the development and implementation of performance measures are often viewed as managerial control devices and solely for the benefit of management. As a result, employees often tend to respond with distrust to the implementation of performance measures in their workplace. Performance measures can become a source of division and conflict between managers and their employees. It can even result in adverse results wherein employees circumvent intended outcomes.

To create a positive context for performance measures within enterprises at lower stage of maturity, enterprise business executives, managers and leaders must keep in mind that the purpose of a performance measure is to improve from the current of activities state toward a better state. The following are Dean Spitzer recommendations for creating a positive or improving the prevailing "atmosphere" within the enterprise business in relation to performance measures:

1. *Start on a small pilot scale.* Start the development and implementation of performance measures wherever possible. Of course it is desirable to be able develop and implement performance measures through the entire enterprise business, but when that is not possible, start from a secure position from which further progress may be made.
2. *Emphasize the improvement and learning purposes.* Ensure that those involved understand and absorb that the development and implementation of performance measures must be focused on improvement and learning purposes (which are the highest and most motivating purposes of a performance measure) to achieve the enterprise intended strategy, otherwise ulterior motives will take over.
3. *De-couple performance measures from judgment and rewards.* Make it clear that the development and implementation of performance measures will not be used to judge, or as the direct basis of rewards. Both are incompatible with the experimental attitude that is essential for the development and implementation of performance measures and receptivity to learning from the data.
4. *Emphasize the importance of honesty – even if the truth hurts.* Subjective performance measures, which do not have a direct impact on realizing the enterprise intended strategy, but are simply the product of a large number of

individual decisions more or less coordinated, are worthless unless respondents feel safe to be honest. After all, if you don't tell your doctor where it hurts, you are not fulfilling your responsibilities as a patient.

5. *Do not let the initiative devolve into another "report the metrics" exercise*. Old habits are difficult to break. There is always the possibility that the development and implementation of performance measures will revert back to another reporting exercise. Be vigilant, do not let that happen!

6.2.2 Focus: Select the Right "Performance Measures"

The second key to making progress on the development and implementation of performance measures is "Focus." It concentrates attention on what is important: aligning performance measures with the enterprise business intended strategy, and with what needs to be managed, relative to the opportunities, capacities, and skills at hand. Within an enterprise business, there is a variety of things on which performance measures that can be defined. But the right performance measures will provide focus and clarity to the management and the remaining employees. The purpose of focusing a "performance measure" is to differentiate between the critical few high leverage drivers of the most important outcome of the enterprise business intended strategy – if these outcome were only known – and the variety of other performance measures – the trivial many – that permeate every area of an enterprise business and keep the enterprise business running.

Defining and developing generic or standard industry-approved performance measures should not be sufficient as these generic performance measures are satisfactory for maintaining the status quo, keeping the enterprise business running, but not for taking the enterprise business to the next stage of maturity and differentiate itself from the competition. In today's highly competitive and increasingly services-oriented marketplace, it is vital that an enterprise business differentiates itself from the competition.

As performance measurement expert Dean Spitzer pointed out, the industry-approved performance measures "*are like your body's 'vital signs' – important, but they won't get you to the pinnacle of health; they are not differentiators (unless the enterprise intended strategy focuses on operational efficiency, and then certain generic performance measures can be highly strategic).*"

Of course, properly defining and developing generic or standard industry-approved performance measures is necessary for winning in the marketplace, but it is not sufficient for sustainable long-term performance.

For example, all companies need to define performance measures on revenue, cost, profit, and customer satisfaction; manufacturing companies need some way to define performance measures on raw materials' costs, productivity, and quality; insurance companies can't stay in business if they don't define a performance measure on risk; banks have to define performance measures on deposits and return on their investments. The list of important but generic performance measures is virtually endless.

In order to thrive – not just survive – and move to a higher stage of performance measurement maturity, enterprise business executives, managers and leaders need to focus attention on the few critical performance measures that matter most to really drive the enterprise business performance, and focus all employees' attention on those performance measures. The following are adapted Dean Spitzer's recommendations for defining and developing focused performance measures:

1. *Challenge old and outmoded assumptions.* Be willing to challenge the existing mental performance measures models. This requires that the implicit (and often hidden) assumptions behind existing performance measures be made explicit.
2. *Take time to understand the breakthrough desired.* There is no substitute for taking the time to understand the situation that has given rise to the need for new performance measures and what is expected of the new performance measures.
3. *Make sure any new performance measure is linked to the enterprise intended strategy.* All new performance measures must reflect the performance goals of the enterprise as a whole. They should be clearly related to what the enterprise wants to accomplish and how the enterprise does business. Determine how the new performance measure will help release more of the enterprise unused strategic potential.
4. *Make sure that the performance measures are doing what they are supposed to do.* Make sure they are yielding valuable knowledge, driving the right improvements, and not having negative side-effects.

6.2.3 Integration: Align "Performance Measures"

Integration addresses the flow of measured information throughout the enterprise business so that the co-variations of different performance measures can be observed relative to the overall value created. Focusing on isolated functional performance measures causes sub-optimization, which is improvement of isolated functions to the detriment of overall organizational effectiveness, and tends to build functional "silos"[1] that focus exclusively on their own self-serving measures. For example, too much focus on profitability can actually undermine customer loyalty, while too much emphasis on customer loyalty can undermine profitability.

This third key to making progress on the development and implementation of performance measures is the overall trade-offs and balance among several different factors to create an optimal configuration of performance measures across the enterprise business. The objective of integrating performance measure is to avoid cross-purposes operations within an enterprise business. Individual performance measures can be poorly used if they are not integrated into a larger "measurement framework" that shows how each performance measure is related to other important

[1] Term used to denote areas within an enterprise where managers occupy a privileged position in terms of resources and influence, and where they use this for their own, self-interested, functionally-oriented motives rather than for the wider benefit of the enterprise.

measures, and how the constructs (which the performance measures represent) combine to create value for the enterprise as a whole.

There are two types of integration: vertical and horizontal. Vertical integration involves the connection between an enterprise business intended strategy and performance measures up and down within the enterprise business units and/or departments. Horizontal integration is the connection of performance measures across the enterprise business units and/or departments. Aligning what employees are currently doing within the enterprise business is not the key to successful integrated performance measures, but aligning what employees should be doing – executing the enterprise business intended strategy – as we will illustrate in a next chapter.

6.2.4 Interactivity: Develop Dialog on "Performance Measures"

The fourth key to making progress on the development and implementation of performance measures is "Interactivity." Interactivity speaks to the inherently social nature of the purposes of performance measures, so that it embodies an alignment with the enterprise business model, intended strategy, and operational imperatives. It represents the social communicative aspects of "performance measures" which occurs through a search for shared meaning or understanding. Because performance measures need to be integrated across an enterprise business, functions and people affected by those performance measures within the enterprise business need be become more interactive through dialogue.

This interaction around them is what will turn the development and implementation of performance measures within the enterprise business into a changed and effective reality. Unfortunately, very few people are skilled at dialogue, and very few enterprises business currently have a strong capacity for dialogue. In fact, in most enterprise businesses dialogs are suppress in favor of debates, the more formal and adversarial processes which are antithetical to dialogues, because the purpose is for one individual to win an argument.

In order to take advantage of the interactivity that should occur at every stage of the development and implementation of performance measures, enterprise business executives, managers and leaders must endeavor to create a positive context of performance measures. Dialogue thrives on openness, honesty, and inviting multiple viewpoints. In dialogue, diversity of perspective is almost always good – whether it be functional, cross-functional, local, global, systemic, etc. The more perspectives involved, the richer the dialogue can be around performance measures. In relation to performance measures within an enterprise business, dialogue as interactivity should incorporate: learning, understanding, defining, listening, modeling, hypothesizing, balancing, linking, and integrating.

Although most enterprise businesses have a long way to go on the development and implementation of performance measures, enterprise businesses at the "Continuous Improvement" stage of maturity have been identified as being more effective than most in using performance measures. They have enviable records of both financial and nonfinancial performance. These enterprise businesses at the "Continuous Improvement" stage of maturity are not more successful because they use isolated performance measures, but because of how much more effectively they use performance measures as a critical part of managing and doing work on a continuing basis.

6.3 Assessing Performance Measurement Maturation

Spitzer takes a developmental approach to measurement improvement, providing a Measurement Maturity Assessment. Clearly, the transformative potential of performance measurement is dependent on the maturational complexity of the context in which it is implemented. Spitzer outlines the ways in which each of the four keys plays into or hinders transformation and maturation. He also provides practical action plans and detailed guidelines.

Alternatively, the survey form in Table 6.1 can be used to assess the current status of performance measures maturation within an enterprise business.

Table 6.1 Enterprise performance measure context questionnaire

#	Observation	Rating
01	I am confident that that the most important factors for the present and future success of the enterprise are being effectively measured	
02	Performance measures are continuously aligned and re-aligned with the enterprise intended strategy	
03	The importance and value of performance measures are acknowledged throughout the enterprise?	
04	Employees in the enterprise proactively seek and welcome performance measurement-based feedback	
05	Employees have positive attitudes toward performance measures, trust it, and are confident that it will not be used against them	
06	Performance measures in the enterprise are timely and easy to understand	
07	Performance measures are being routinely converted into knowledge and insight for learning purposes	
08	Projects and initiatives in the enterprise are measured for effectiveness (not just for cost and timely completion)?	
09	The enterprise performance measures foster decisiveness, openness, transparency, and collaboration	
10	The enterprise performance measures foster cross-functional collaboration	
11	The enterprise stakeholders understand the cause-and effect relationships, dependencies, and trade-offs among key performance measures	
12	The enterprise stakeholders feel confident that the performance measures are providing the insight and foresight to guide high quality decision-making	
13	Significant progress is been made to integrate performance measures (especially those related to customers) to enable more decision making	
14	Continuous improvement of the performance measures framework and updating performance measures is an enterprise priority	
15	Progress is being made in setting performance measures on difficult-to-measure sources of intangible value (e.g., talent, knowledge, innovation)	
16	The enterprise is open to experimentation with new, innovative, and cross-functional performance measures	
17	There are frequent interactivity and positive dialogues about performance measures in staff and management meetings	
18	The enterprise performance measures framework is dynamic and flexible enough to adapt quickly to increasing complexity and changing circumstances	
19	Are performance measures used at least as frequently for improvement and learning as it is for monitoring, reporting, and rewarding	
20	There is a major effort in the enterprise to educate employees about performance measures	
Use the standard five-point rating scale:		
5 = Strongly agree; 4 = Agree; 3 = Neither agree nor disagree; 2 = Disagree; 1 = Strongly disagree		
Interpretation key:		
Highest score is 100		
High score range is 75–100. Your enterprise business is doing a good job on performance measures		
Moderate score range is 50–75. Your enterprise business is making progress on performance measures		
Low score is below 50. Performance measures framework in your enterprise business is still very much in need of improvement		
Danger zone is below 25		

Performance Management

7

Performance management is the comprehensive set of activities followed to establish, implement and improve an enterprise business performance. It includes defining expectations and accountabilities, setting performance standards and performance measures, and assessing results. It is the centralized and coordinated management of performance measures to: obtain the benefits and control not available from managing them individually and, achieve the enterprise intended strategic objectives and benefits.

7.1 Purpose of Performance Management

Performance management is build and developed within a "performance measures framework" that shows how each performance measure is related to other important measures within the enterprise. A performance measures framework itself is a more comprehensive hierarchical organization of performance measures that fit together according to a logical structure. A good performance measures framework provides enterprise business executives, managers, leaders and employees with visibility into how their local performance measures fit with the enterprise global performance measures. As such, it provides a "line of sight" that enables enterprise business executives, managers, leaders and employees to appreciate the linkages between what they are doing and what is important to the enterprise as a whole. A good performance measures framework also should not just depict the way an enterprise currently measures itself, but help predict future performance, so that better decisions can be made within the enterprise.

Within a performance measures framework, the purpose of performance management is not about filling out "template" scorecards. But it is about obtaining increasingly deeper understanding that will lead to progressively better actions to drive desired results, and then communicating that understanding throughout the enterprise through integrated performance measures so that everyone can execute in an integrated manner across the entire enterprise.

A. Van Aartsengel and S. Kurtoglu, *A Guide to Continuous Improvement Transformation*, Management for Professionals, DOI 10.1007/978-3-642-35904-0_7,

In other words, the purpose of performance management is to achieve continuously better (and deeper) understanding of how the enterprise intended strategy translates into desired outcomes and drivers of these outcomes, and then to find the best mix of performance measures that conveys strategic intent and integrates the organization to execute that intended strategy. Finding the best mix of performance measures is achieved through tradeoffs decisions. With an overall performance measures framework that shows the relationships between performance measures, it is easier to make the proper trade-off decisions, so that more optimal decisions can be made.

There is no single right way to develop and implement a performances measures framework. Commonly used performances measures frameworks are based on the "strategy map" approach, the four perspectives (Financial, Customer, Internal, Learning and Growth) of the balanced scorecard approach, or on the "performance prism" approach.

7.2 The Balanced Scorecard

The origins of the balanced scorecard go back to a research project at the beginning of the 1990s carried out by Kaplan and Norton (1996). The aim of the project was to describe the essential ingredients of business success by developing a tool that would display, quantify and communicate all the performance measures which are important to a company's success.

The tool developed, labeled the balance scorecard – which achieve widespread acceptance, is a set of performance measures that is balanced by having multiple perspectives or dimensions, including both financial and nonfinancial performance measures within an enterprise. As shown in Fig. 7.1, it balances the financial perspective with customer, internal, and learning and growth perspectives.

Customer Perspective: This perspective is concerned with performance measures which represent the customer expectation. When choosing performance measures for the Customer perspective of the Scorecard, a set of critical and challenging questions must be answered, amongst other:

1. Who are the enterprise business target customers?
2. What is the enterprise business intended strategy in serving them?
3. What do the enterprise business customers value, and what are they willing to pay for?
4. How do we know if we are truly delivering value to our customers?
5. How do we know if we are hitting the mark and dealing with problems before they become customer complaints?

The Customer perspective will normally include attributes of performance measures widely used today: customer satisfaction, customer loyalty, market share, and customer acquisition, for example.

Internal Process Perspective: This perspective of the Scorecard is concerned with performance measures related to the enterprise key processes and activities

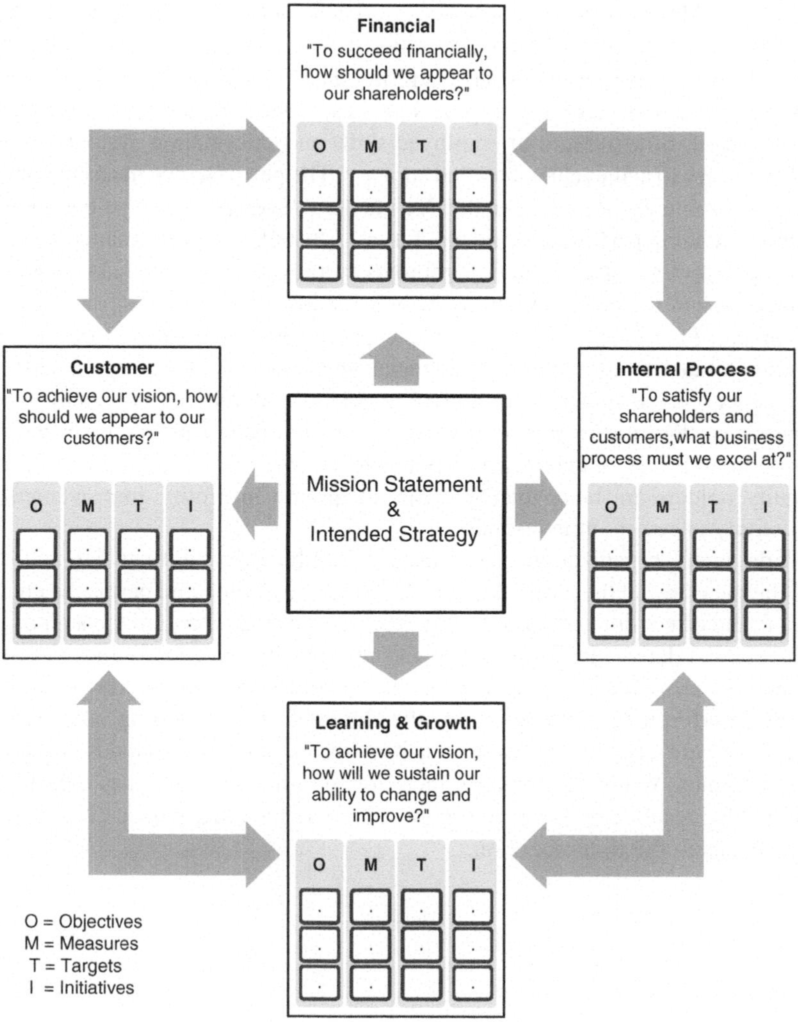

Fig. 7.1 Kaplan & Norton Balance Scorecard's four perspectives

that ultimately drive customer and financial performance. Key processes and activities, which are those having the biggest influence on customer satisfaction and the achievement of the enterprise intended strategy, often include: product development, production, manufacturing, delivery and support. The performance measures here focus on existing operations and will normally include attributes related to increase efficiency, quality, productivity, and reduce cost, and cycle time.

Learning and Growth Perspective: This perspective is concerned with the performance measures related to the capabilities and knowledge that the enterprise must built up to achieve the objectives stated in both customer and internal process perspectives. The performance measures in this perspective are the enablers of the

Customer and Internal Process perspectives. They are the foundation on which this entire Balanced Scorecard is built. Once performance measures and related initiatives in the enterprise Customer and Internal Process perspectives have been identified, there will certainly appear some gaps between the enterprise current organizational infrastructure of employee skills and information systems, and the level necessary to achieve the intended strategy. The performance measures defined in this "Learning & Growth" perspective must be developed to close the gaps and ensure sustainable performance for the future. Attributes of performance measures in this perspective often include: improve employee skills, increase employee satisfaction, and increase availability of information.

Financial Perspective: The financial perspective is concerned with performance measures related to the enterprise financial objectives. If an enterprise correctly implements the customer, the internal process, and the learning and growth perspectives, then the financial perspective will ultimately makes plain the economic consequences of enterprise activities by showing how well the expectations of equity holders are being met in terms of growth in profits, improvements in productivity or return on investment.

While many organizations have used a combination of financial and non-financial measures in the past, what sets the Balanced Scorecard apart is the concept of cause and effect linkages. The four perspectives of the scorecard are supposed to be causally related. Performance measures in the Financial and Customer perspectives are criteria of success on the outcomes that an enterprise wants to achieve; Performance measures in the Internal and Learning and Growth perspectives are criteria of success on how the organization intends to achieve these outcomes. While performance measures on financial outcomes measure the desired "final result," the key is to use the drivers in the other perspectives to move the financials in the right direction.

8 Alignment and Commitment

The attainment of intended strategic goals is the lifeblood of any enterprise business. Ideally, every employee should be acting in concert with the enterprise business intended strategy. The larger the enterprise business is, the more important the synergy that diverse resources and capabilities provide within the enterprise business. No matter how clear and valid an enterprise business intended strategy and its alignment are, if employees at different organizational levels within the enterprise business are not genuinely on the same page, working with each other, rather than against or independent of each other, the enterprise business will not achieve the success that it want. There are three key dimensions to achieving this: the alignment process, the commitment of employees on the resulting alignment, and the development and management of teams to execute the aligned activities.

This chapter delves into the key characteristics and constituents of alignment and commitment necessary to take your enterprise business to the "Continuous Improvement" stage of maturity as described in a previous chapter.

8.1 What Is Alignment?

Alignment relates to "*the degree to which the components of an enterprise business are arranged and focused to optimally support its intended strategy.*" The components of the enterprise business that must be "aligned" include:

1. The performance measures, goals, skills and capabilities, and the hearts, minds, and behaviors of both the people doing the work and organizational leaders;
2. The work being completed (i.e. operational work, key business projects and processes); and
3. The plans (e.g. learning and development/human capital plans), tools and technologies, and resources that support the work being completed.

Building and ensuring alignment within an enterprise business requires focused action and is an ongoing activity, the outcome of which must be continuously assessed and improved.

A. Van Aartsengel and S. Kurtoglu, *A Guide to Continuous Improvement Transformation*, Management for Professionals, DOI 10.1007/978-3-642-35904-0_8,

Fig. 8.1 The "rowing eights" metaphor for enterprise alignment

Let's refer to Fig. 8.1, which we call the "rowing eights" metaphor for alignment. This metaphor is rich with meaning. Think about the performance of the top teams at the Olympics – what do you see? Look closely and you will see nine people (including the coxswain) working together in synchrony and commitment to achieve their primary goal of crossing the finish line first.

In the boat race, the coxswain is a vital addition to the crew. Besides steering, the coxswain is also the voice in the boat, coaxing, motivating, and calming an eight-individual engine. As the coxswain faces forwards, in a conventional boat, stroke side (even numbers) is to the left and bow side (odd numbers) to the right.

It does not take long before the stroke oarsman – even one who has trained for 6 h a day for 6 months – feels fire in the lungs and the legs. International races last, at most, 6 min, so a race three times as long is cruel and unusual punishment. The individual who can best stand that punishment is the stroke, setting a pace that perhaps only he knows the crew will be able to maintain throughout the race. It is important, though, that the rest of the crew is able to stay with him and, over a course of this length, finding a steady rhythm is vital.

"Seven," the translator for the bow side of the boat, takes Stroke's rhythm and acts as the individual to follow for the blades behind his. If "Seven" does not back-up the stroke's commitment or follow any change in pace, it is certain that none of the rest of the crew will. Stroke and Seven make up the Stern Pair.

The middle four of the boat are the engine room, the biggest and most powerful members of the crew. But, of the four heavies, "Six" is the brains of the operation, making sure that the rhythm of the stern is not lost when it reaches the less subtle middle of the boat. Often one of the tallest members of the crew, "Six" lends length to the mix – the longer the stroke, the further the boat will travel.

The main demand to the individuals in the middle of the boat is to provide power – as much as possible. Another member of the powerhouse, "Four" has to remember that there are three individuals behind him who are a long way from the action in the stern, and must help to keep them in touch.

In lesser boats, the "Three" seat is where the least technically able oarsman sits. This individual is not far enough towards the stern to upset the rhythm but, as he/she is not right at the bow, any error will not result in the boat swerving. At boat race level there are no weak links, but "Three" still has the least responsibility.

"Two," often the seat occupied by the back-up stroke, joins the individual behind him to make up the Bow Pair who, as the first blades to catch the water at the front of the boat, must be the sharpest members of the crew at the beginning of the stroke.

As the man at the front of the boat, "Bow" joins two in making sure that the boat is balanced correctly. His blade makes the most difference when it is placed in the water, so he must be sharp and technically correct. Anyone can do that at the start of a race but, after 18 min, especially in a close contest, only the best will be able to hold the correct shape throughout.

With all this in place, teams power their boats on a straight course, translating 100 % full commitment of their effort into the forward movement that moves them closer to their goal. However, when team alignment and cohesion is off, the boat strays off course, essentially wasting time, energy, and the resources that were invested in trying to achieve the goal of winning the race.

Successful rowing eights operate as a cohesive unit, moving together so as not to upset the balance of the boat and slow it down. To achieve success, the rowers must stroke at the same pace with the blades of every oar pulling at the same depth in the water. They all know the overall game plan for success and they are committed and ready to respond to the orders of the coxswain as individuals and as a cohesive unit. Each member of the team knows what their task is during the race and that they can rely on their coaching, training, boat, and equipment, and the skills, technique, and commitment of their teammates while the race is on.

It is very much the same for an enterprise business. Without alignment and team management, the best enterprise business intended strategic plan will never be fully achieved because alignment and team management is the glue that makes possible realization of an intended strategy execution excellence within an enterprise.

Indeed, most enterprise businesses are composed of pieces vying for scarce resources operating more like competitors than cooperators – acting individually, without regard to systemic interdependencies. These enterprise business components and their constituent people do not act this way because they are obstinate or nasty. They do it because the enterprise business systems condition them to do so. They are simply following the traditional, if flawed, logic, which is: "If every function within the enterprise business meets its goals ... if every function hits its budget ... if every project is completed on time and on budget ... then the enterprise will win." On the other hand, it should be clear that such thinking no longer works, if it ever did.

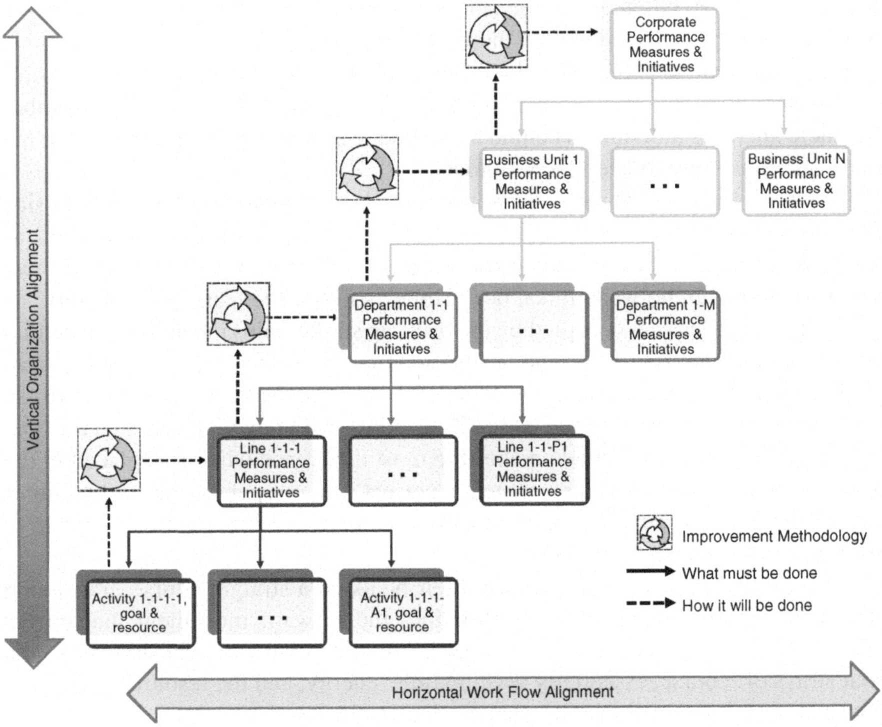

Fig. 8.2 Vertical and horizontal alignments

An enterprise business alignment focused on the performance of the whole by integrating organizational and work flow components of the enterprise business vertically and horizontally as illustrated in Fig. 8.2.

Vertically, alignment relates to the integration of organizational components up and down within the enterprise business units. It entails communicating and deploying the enterprise business intended strategy, in a structured form, from the highest organizational level of the enterprise business down to activities at the lowest organizational level within each business unit of the enterprise business. Horizontally, alignment relates to the integration of work flow components across the enterprise businesses.

Components of an aligned enterprise work together as a unit in a new way (i.e. more efficiently, effectively, and dynamically) on the right things that achieve the intended strategy and deliver the greatest business value. An aligned enterprise business gets things done faster, with less effort, and with better results, and is more agile and responsive to changing business conditions. A high degree of alignment is indicative of a consistent behavior and performance across the enterprise business. A very low degree of alignment is representative of enterprise business composed of a collection of functional silos that operate so independently that there seems to be little connection among them at all and a proliferation of projects.

8.2 How to Realize Alignment?

Building an enterprise business alignment is not an easy task and there are several approaches to perform alignment of the components of an enterprise: Strategy Maps, Balanced Scorecard, Policy Deployment (Hoshin Kanri), or Quality Function Deployment. The approach that we advocate in the following section follows the Quality Function Deployment constructs.

The Quality Function Deployment (QFD) methodology is widely used in engineering for systematic and effective planning (Xie, Goh, & Tan, 2003; Madu, 2000; Akao, 2004; Terninko, 1997; ReVelle, Moran, & Cox, 1998; Cohen, 1995; Bossert, 1991; Ficalora & Cohen, 2010; Duffy, Moran, & Riley, 2010; Day, 1993; Bhardwaj, 2010). The designation "Quality Function Deployment" does not immediately give an understanding of what the methodology is or does. This designation comes from the original Japanese phrase consisting of three characters, each of which has several meanings:

1. *Hin shitsu*, which can mean "quality," "features," "attributes," or "qualities"
2. *Kino*, which mean "function" or "mechanisms"
3. *Ten kai*, which can mean "deployment," "evolution," "diffusion," or "development"

Basically, QFD means deploying the attributes of a product or service desired by the customer throughout all appropriate functional component of an enterprise business. It also provides a mechanism for its achievement, that is, the set of relations (in the form of matrices) that serves as both a structure and a graphic of the planning deployment process. QFD was developed by Yoji Akao in Japan in 1966. By 1972 the power of the approach had been well demonstrated at the Mitsubishi Heavy Industries Kobe Shipyard (Sullivan, 1986) and in 1978 the first book on the subject was published in Japanese and then later translated into English in 1994 (Mizuno & Akao, 1994).

Our need for using a QFD approach to enterprise business alignment is driven by two related objectives, which start with the enterprise business intended strategy and ends with the resources that support the work being completed to realize the intended strategy. These two objectives are:

1. To convert the enterprise business intended strategy into substitute intended strategy characteristics for use at operational stage;
2. To deploy the substitute intended strategy characteristics identified at the operational stage to the line activities, thereby establishing the necessary control points and check points.

8.3 Enterprise Business Alignment Process

Developing an enterprise business alignment is a complex task requiring involvement and commitment from individuals in a number of different functional areas within the enterprise. For this reason, it is useful to see the enterprise business alignment as a project.

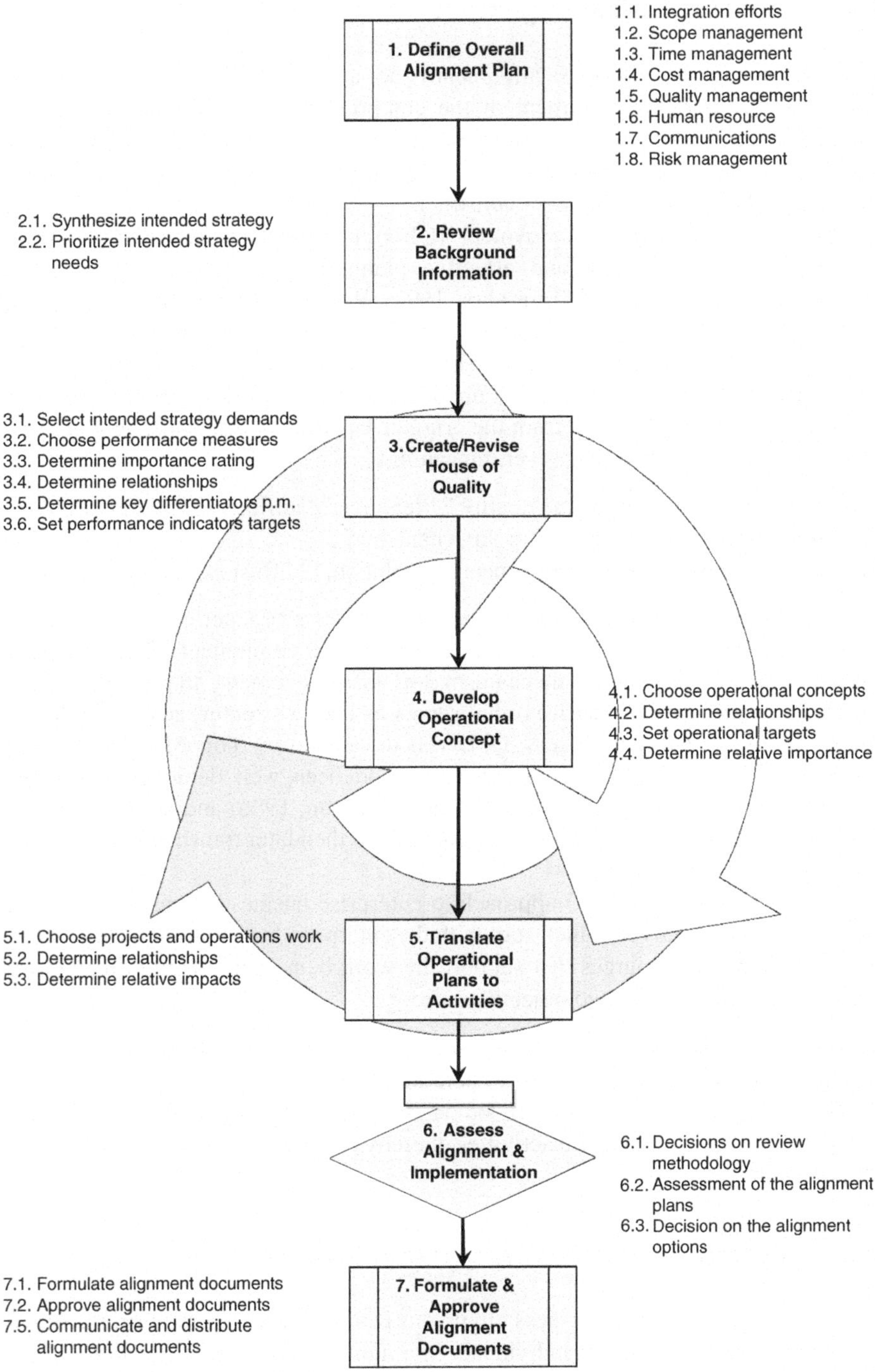

Fig. 8.3 Enterprise alignment process

Within a project management context, the alignment process, illustrated in Fig. 8.3, includes the actions necessary to define, analyze, develop, implement, assess and formulate all subsidiary plans into alignment documents.

8.3.1 Define Overall Alignment Plan

As with any project, the initial step of the alignment process consists of activities that integrate the various elements within the alignment process. This integration includes characteristics of unification, consolidation, articulation, and integrative actions that are crucial to completion of the alignment, successfully meeting customers and stakeholders requirements, and managing expectations. Within this step, choices are made about where to concentrate resources and effort on any given day, anticipating potential issues, dealing with these issues before they become critical, and coordinating work for the overall alignment planning good.

The main objective of this initial step is to effectively integrate the activities that are required to accomplish the alignment objectives within the enterprise defined procedures.

8.3.2 Review Background Information

The second step of the alignment process aims to gather, review and integrate background information needed to establish the various relations to be used in the conversion of enterprise business intended strategy to substitute strategy characteristics. Each member of the alignment project team will approach the project with certain preconceived notions regarding the nature of the business, its competitive position, future prospects, appropriate intended strategy, and performance measures.

Within this step, team members must gather and review as much background material as they can find. Each member of the alignment project team has a particular background and experience, but to build an effective alignment, they must have access to the total pool of information that exists on the enterprise.

Acquiring this information typically requires input from marketing sources (internal and external), technical sources (internal and external, in some instances including suppliers of some state-of-the-art components), field service personnel, customer service representatives and, most importantly, representatives from several different sets of key customers. Here are some of the sources of information to consider:

1. Financial Perspective
   - Annual report
   - Performance reports
   - Analyst reports
   - Trade journals
   - Benchmark reports

2. Customer Perspective
   - Marketing department
   - Trade journals
   - Consulting studies
   - Project plans
   - Strategic plan
   - Performance reports
   - Benchmark reports
3. Internal Process Perspective
   - Operational reports
   - Manufacturing reports
   - Competitor data
   - Benchmark reports
   - Trade journals
   - Consulting studies
   - Project plans
4. Learning and Growth
   - Human resources data
   - Trade journals
   - Core values
   - Benchmark reports
   - Consulting studies

The task of unearthing background material during this step of the alignment process aims specifically at discerning the current degree of consistency of activities within the enterprise. It also aims to (a) synthesize the enterprise intended strategy demands and then (b) prioritizing those needs.

8.3.3 Determine Key Differentiator Performance Measures

The third step of the alignment process is concerned with the translation of enterprise business intended strategy demands to substitute performance measures, language and priorities. The translation is done by building the "House of Quality" associated with the enterprise intended strategy demands and performance measures.

Figure 8.4 shows a typical "House of Quality" matrix, which is the foundation of all QFD exercises. The matrix structure and visual nature of the "House of Quality" give both discipline and guidance to the conversion process by exploring the information that it contains.

The demands, represented on the left side by WHATs – and the performance measures – represented on the top by HOWs – are the input to the matrix and the foundation for further activities. In some instances, the alignment project team may be able to use these inputs to identify new concepts that represent opportunities to be exploited to gain competitive advantage.

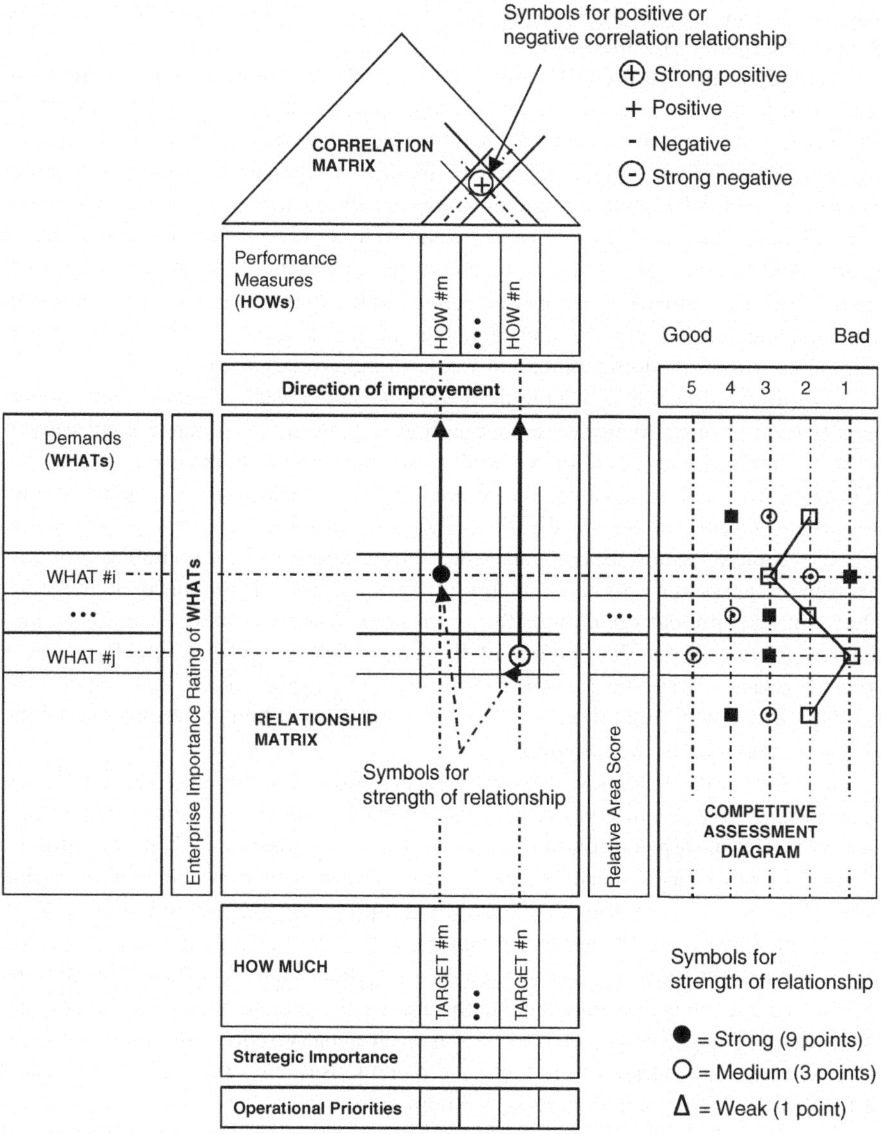

Fig. 8.4 The house of quality

Next to the demands is the importance rating, where the current intended strategy demands stand "from the enterprise business perspective." It is important to emphasize to the alignment project team that "this is how the enterprise business feels." The importance rating can use any scale. Traditionally, the importance rating is scaled from 1 (low) to 5 (high). On some projects, this scale is changed to 1 (low) to 10 (high) when a finer breakdown is needed. During the selection scale for the importance rating, some team members of the alignment project might state that "everything is important." In the purest sense, that is true, but there is also a hierarchy of importance.

The body of the "House of Quality"; i.e., the "Relationship Matrix" is where the relationships are categorized. This is where intended strategy demands "translated" into operation terms. It is also where interactions between relationships between a given WHAT and a given HOW are identified so that the synergistic effect meeting the intended strategy demands is seen; by meeting one intended strategy demand, other demands can also be satisfied. In filling in the "Relationship Matrix," a great deal of dialogue will take place between members of the alignment project team. Here the team will identify the relationships between the intended strategy demands and the performance measures that matter the most. The relationships are defined as strong, medium/some, and weak/possible relationships.

A strong relation equals 9, a medium equals 3, and a weak equals 1. Symbols are usually used in order to aid in the recognition of patterns. Numbers are substituted in at a later time to calculate weight at the bottom of the matrix. During the dialogues that will take place, the alignment team members must address both the operation content (especially the validity of the intended strategy demands) and the context issues. Specifically, they must address the inevitable concerns (spoken or unspoken) about how this effort will differ from those of the past. They must keep clear communication with managers from their respective function/department about the content of the intended strategy and directly address context issues. They must ensure that these managers clearly understand the validity of the new alignment – its content – and how their functions contribute to it and support it in the long term.

The "Correlation Matrix" ("Roof" of the House) compares the HOWs to determine if they are in conflict or assisting each other. It identifies positive and negative relationships/correlations, that is, technical trade-offs. This is valuable because in most cases, these trade-offs have not been documented prior to this time. The trade-offs are often the source of compromises because of the limitations of currently available resources or facilities. By identifying them early on, the alignment project team can narrow their efforts. Although trade-offs and limitations are fact of life, they do represent opportunities for breakthrough improvements. Once a strong negative correlation is broken, it usually represents a major model shift resulting in proprietary products and patents. Basically the roof of the House of Quality represents a hotbed of opportunities.

The right side of the matrix illustrates the "Competitive Assessment Diagram." It shows how the key WHATs can be combined with the importance rating and competitive benchmarking. This is where the competitive analysis and the sales point are listed. The sales point is a measure of how sellable a particular intended strategy demand is. If an enterprise business is considered the "best," it is a high sales point and should be included if not already in the enterprise sales literature and training. If it is not considered the "best," it should be a low sales point. Enterprise business executives, managers and leaders should promote the enterprise best intended strategy demands. The "Competitive Assessment Diagram" column is used to develop marketing strategies. It displays not only strength and weaknesses of WHATs, but also opportunities for breakthrough.

It is imperative that the same scale be used here as was used for the importance rating of intended strategy demands.

Below the "Relationship Matrix," the alignment project team sets strategic priorities and expected levels (HOW MUCH) of performance indicators associated to the selected performance measures. These expected levels form the "wish list" that drives the alignment effort. Some compromises may be required when all of the wish list items are not attainable. The decision here is to improve intended the strategy, remain equal to the competition strategically, or remain behind the competition. Improvement is desired in most enterprises, but may not be attainable if the competition is considered best in the marketplace. In these cases, parity may be the only option, unless due to some constraints the enterprise has to take a lesser position. Reality sometimes forces this decision, so the alignment project team must be made aware of constraints to the intended strategy if there are any.

It is important to note here that the alignment project team is now working with more objective language than the original subjective language of the WHATs. From this information and from the competitive analysis, a planning weight will be calculated. This planning weight identifies the performance measures that are the most critical for success as well as the degree of technical difficulty to achieve. It will help the team focus on the performance measures that will yield the greatest potential for success in the marketplace.

It is sometimes desirable to add a few more rows to the bottom of the matrix. These rows will indicate:

1. The scale-up, which is the ratio of the expected level to the current level. This is an indicator of how much effort is needed to meet the expected level. The higher the number, the more effort is needed. This number must be looked at carefully.
2. How difficult it will be to get to the expected levels of performance indicators associated to the performance measures.
3. How long the alignment effort will take to reach the expected level and
4. Whether the enterprise is going to pursue the expected level (improvement) or merely the current level (parity).

In this third step of the strategic planning process, the WHATs are obtained from a synthesis of the enterprise intended strategy.

Figure 8.5 shows an example of complete initial "House of Quality" of an enterprise business unit pursuing a cost reduction or efficiency, and operation and process excellence intended strategy.

In this example, the performance measures developed by the alignment project team represent factors that mattered most to the business unit. As such, the best trade-off on these performance measures is no trade-off, as indicated by the empty correlation matrix. The weighting scheme for the importance rating of intended strategy demands is increase (resp. decreasing) from 1 to 5 (resp. 5 to 1), 1 (low) and 5 (high).

In order to set clear priorities, the values of symbols for strength of relationship are chosen to place a high emphasis on strong relationships. The overall score of an intended strategy demand (i.e. area score) is the sum of values of symbols

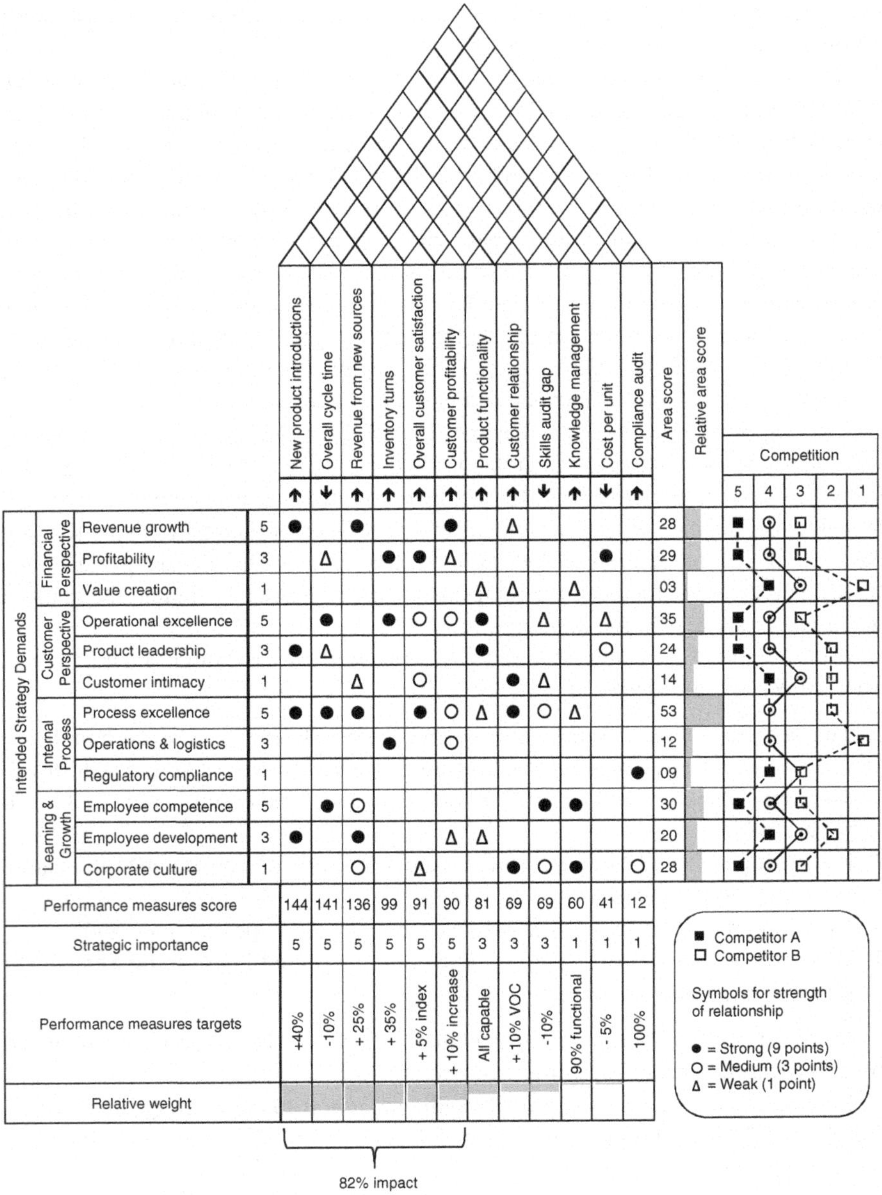

Fig. 8.5 Translating intended strategy demands into key differentiator performance measures

for strength of relationship of performance measures affecting that intended strategy demand.

This overall score provides an indication on how well the performance measures that matters the most measure the intended strategy demand. A low value indicates that the intended strategy demand is less measured by the selected performance measures.

The overall score of a performance measure is the weighted sum of values of symbols for strength of relationship of intended strategy demands being measured by the performance measure. The weights are given by the enterprise importance ratings of intended strategy demands. This overall score of a performance measure shows the impact of the performance measure on the intended strategy demands.

To set further priorities to achieve the intended strategy and stand out in the marketplace for customers and stakeholders, the alignment project team has set strategic importance weights (5 = key differentiator, 3 = differentiator, 1 = key requirement) on performance measures.

The overall score of a performance measure weighted with these strategic importance weights is put into a Pareto chart to show which performance measures are most important in meeting the intended strategy demands. The resulting performance measures, the relative overall weighted sum of which is around 82 %, are the performance measures on which the enterprise business unit will focus on and allocate resources throughout the intended strategy time period. The remaining performance measures will be treated separately and will not be driven to new heights, but will be maintained at their current levels.

8.3.4 Determine Operational Performance Concepts

The fourth step of the alignment process is concerned with the translation of key differentiator performance measures to operational performance concepts, language and priorities.

This translation is especially important because it typically takes the key differentiator performance measures demands language and turns it into technical language with which employees at different organizational levels within the enterprise can work.

The translation is done by building the "House of Quality" associated with the key differentiator performance measures and the operational performance concepts that matter the most, and which can be measured.

In this second "House of Quality," the key differentiator performance measures (WHATs) are input to this matrix. The performance measures score associated with these key differentiator performance measures are the importance ratings. The operational performance concepts that matter the most (HOWs) designed to support achievement of the intended strategy demands are also input to this matrix.

The construction of this second matrix follows the same sequence used to build the initial matrix, with a more abbreviated set of steps since the alignment project team is now narrowing the focus to the critical areas for work within the enterprise business unit. Figure 8.6 shows an example of complete second phase of the "House of Quality" for the enterprise business unit considered above.

When looking at the operational concepts, two techniques can be used to select the ones that matter the most among a large amount of proposed operational concepts. These techniques are:

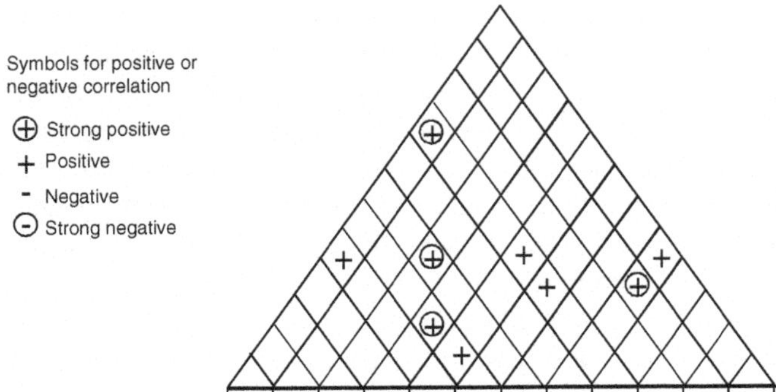

Key differentiator p.m.		Improve agility	Increase direct marketing	Identify new markets	Increase processes clarity	Reduce concept-to-design cycle time	Faster components development	Increase use of lighter materials	Reduce defects and rework	Faster ramp-up to assembly	Integrate press & body shops	Increase in-line quality control	Improve central storage & maintenance	Area score	Relative area score
		↑	↑	↑	↑	↓	↑	↑	↓	↑	↑	↑	↑		
New product introductions	144	○		●		●	●	●	●	●	●	●		75	
Overall cycle time	141	○			●	●	●			●	●	○	○	54	
Revenue from new sources	136	●	○	●			●	●		○	Δ			43	
Inventory turns	99				●				Δ	Δ	○		●	23	
Overall customer satisfaction	91	●						○	●			○		24	
Customer profitability	90	○	Δ	Δ					○	Δ				09	
Operational concepts score		3168	498	2610	2160	2565	3789	2793	2484	3162	2998	1992	1314		
Operational concepts numerical relative score (x 100)		10.73	01.69	08.84	07.31	08.69	12.83	09.46	08.41	10.71	10.15	06.74	04.45		
Operational department id.		1	1	1	2	2	2	3	3	3	4	4	4		
Operational concept targets		+30%	+35%	+ 5%	+ 35%	4 weeks	9 weeks	+15%	-10%	2 weeks	90% functional	+60%	100% clean		
Performance measures relative weight															

Operational department id.: 1 = Marketing, 2 = Engineering, 3 = Manufacturing, 4 = Final Assembly

Fig. 8.6 Translating key differentiator performance measures into operational concepts that matter the most

1. Stuart Pugh's Concept Selection Technique, and
2. The Value Analysis Technique.

Stuart Pugh's Concept Selection Technique – Stuart Pugh's concept selection starts with the premise that the selection of an operational concept that matter the most to support realization of the intended strategy demands is more difficult than the selection of a wrong operational concept, due to a state known as "conceptual weakness."

Conceptual weakness occurs in two ways: (1) a weak operational intend in general and (2) a strong operational intend, but one that is not well thought out so it is subject to debate, which leads to lesser operational intends being chosen.

Pugh's approach is to compare all possible operational proposals with the same level of detail in a matrix format. The top of the matrix has the proposed operational concepts, and the side has the criteria all the operational concepts will be evaluated against. The evaluation criteria include cost, risk, and resources. Included in the top of the matrix is the current operational standard as well as description of each operational concept. The scale used is simple: + means better than current operational standard, – means less than current operational standard, and = means the same as current operational standard.

At the bottom of the matrix is the total for each of all scales (+, –, and =) for each operational concept. If more than one operational concept matter the most, then Pugh's approach looks at the concepts that matter the most with the sign + rows removed. This is repeated until one operational concept emerges. If one operational concept does not emerge, then the operational standard must be changed and the operational concepts re-evaluated.

An assessment of consistency is performed by taking the strong operational concept and resuming the matrix to see if any change occurs in the results. This accomplishes a number of things: better insight into the key differentiator performance measures, a better understanding of the intended strategy demands, better ideas of alternative solutions, the identification of potential interactions, and an understanding of why one operational concept matters more than another. This pattern can be repeated as long as necessary to obtain the level of detail necessary to develop the key differentiator performance measures.

The Value Analysis Technique – Value Analysis (or Value Engineering) is another approach that can be effectively utilized. It can be used to identify the operational concept that adds value and to provide realization at the lowest total cost. Value is determined by looking at both the positive and negative aspects of each operational concept. A typical approach is to have the alignment project team understand the purpose of each operational concept. Then each operational concept is evaluated by a cost-benefit factor, which is plotted on a graph that has importance on the y-axis and cost on the x-axis. A 45-degree line is drawn, and items below that line are targeted for improvement. These targets can then use Pugh's concept selection for determining an alternative operational concept. The new operational concepts are then evaluated for technical feasibility, cost, risk, resources, quality, etc... The operational concepts that matter the most are then chosen to be implemented.

8.3.5 Select Projects and Operations

The fifth step of the alignment process is concerned with the translation of operational concepts that matter the most to execution plans and selection of projects. This translation is very important as it typically takes the operational concepts that matter the most and turns it into (production) line activities, thereby establishing the necessary control points and check points for realizing the enterprise business intended strategy. The translation, as the previous two, is done by building the "House of Quality" associated with the operational concepts that matter the most and the (production) line activities which can be measured.

In this third "House of Quality," the operational concepts that matter the most (WHATs) are input to this matrix. The numerical relative scores associated with these operational concepts that matter the most are the importance ratings. The (production) line activities (HOWs) designed to support realization of the operational concepts are also input to this matrix. The construction of this third matrix follows the same sequence used to build the initial and the second matrices, also with a more abbreviated set of steps since the alignment project team is now narrowing the focus to the critical areas for work at the organizational lowest level within the enterprise business unit. Figure 8.7 shows an example of complete third phase of the "House of Quality" for the enterprise business unit considered above.

In this fifth step, the (production) line activities designed to support realization of the operational concepts include projects and operations activities that matter the most. We should recall that the objectives of projects and operations are fundamentally different. A project is a sequence of unique, complex, and connected activities having one goal or purpose and that must be completed by a specific time, within budget, and according to specification. It is a temporary effort undertaken to create a unique product, service, or result. The purpose of a project is to attain its objectives and then terminate. Projects are therefore utilized as a mean of achieving an enterprise business intended strategy. They conclude when their specific objectives have been attained.

Operations activities are ongoing and repetitive efforts, the purposes of which are to sustain the enterprise business. When their objectives have been attained, operations activities adopt a new set of objectives and the work continues. Although projects and operations activities sometimes overlap, both share the following characteristics: they are constrained by limited resources; they are selected following analyses of their added value in terms of costs and benefits to the enterprise business; they are performed by people; and they are planned, executed, and controlled.

All enterprise businesses are faced with the basic problem of allocating limited resources to many different uses such as current execution of projects and operations activities. These different uses of resources are the means by which an enterprise business can organize its resources in the pursuit of its intended strategy. As indicated in a previous section, aligning the resources on what people are currently doing within the enterprise business is not the key to successful alignment, but aligning the resources on what employees should be doing – executing the enterprise business intended strategy.

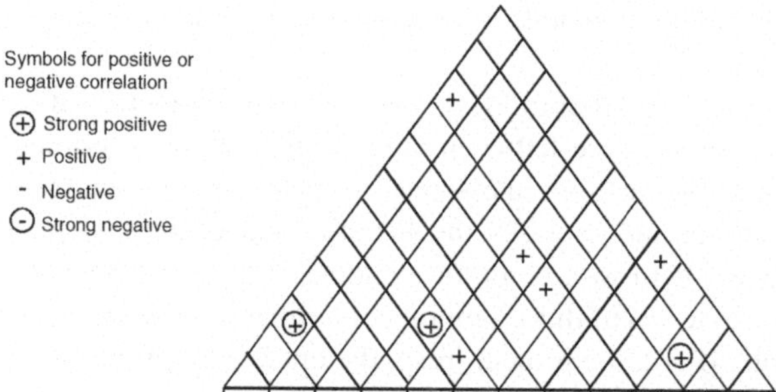

		Operational concepts	Weight	Reducing customer bid cycle time ↓	Reducing customer complains ↓	Implementing marketing analytic and product allocation model ↑	Remanufacturing fit-for-use parts ↑	Improving body welding cycle time ↑	Improving subassemblies transfer process ↑	Reducing work-in-progress inventories ↓	Implementing in-line product self-testing ↑	Reducing assembly throughput time ↓	Reducing component bid cycle time ↓	Reducing parts design time ↓	Reducing prototype production cycle time ↓	Area score	Relative area score
Operational concepts	1	Improve agility	0.107	●	○	○				●		●	●	●	●	60	
		Increase direct marketing	0.016			●										09	
		Identify new markets	0.088			●										09	
	2	Increase processes clarity	0.073				○					Δ	○	○	○	13	
		Reduce concept-to-design cycle time	0.087										●	●	●	27	
		Faster components development	0.128	○		Δ	●	●	●		○	Δ	●	●	●	62	
	3	Increase use of lighter mat	0.095				●	●								18	
		Reduce defects and rework	0.084		●		●							○		21	
		Faster ramp-up to assembly	0.107				○	Δ	●	●			Δ	Δ		24	
	4	Integrate press & body shops	0.102							○		●				10	
		Increase in-line quality control	0.067							●	●					18	
		Improve central storage & maint.	0.045			Δ	○		○							07	
(production) line projects/activities impact score				1.350	1.079	1.442	3.437	2.113	2.252	2.841	0.992	2.080	3.228	3.481	3.121		

Fig. 8.7 Translating operational concepts into production line activities

Examining the availabilities of resources with or without a project (resp. the operation) is the basic method of identifying its costs and benefits. Using limited resources in one direction reduces the resources available for use in another direction. Pursuing of one intended strategy demand may involve a sacrifice in the other objectives. Thus, there are clearly tradeoffs in using resources: a choice has to be made among competing uses of resources based on the extent to which they help the enterprise achieve its intended strategy. If an enterprise business consistently chooses allocations of resources that achieve most in terms of these intended strategy demands, it ensures that its limited resources are put to their best possible use.

Project (resp. Operation) economic and financial analysis is a method of presenting this choice between competing uses of resources in a convenient and comprehensible fashion. In essence, the analysis is based on economic and financial considerations and assesses the benefits and costs of a project (resp. operation) and reduces them to a common denominator. If the benefits exceed costs – both expressed in terms of this common denominator – the project (resp. the operation) is acceptable. If not, the project (resp. the operation) proposal should be rejected. The definition of benefits and costs, however, is such that these factors play an integral part in the decision to accept or reject a project (resp. the operation).

Benefits are defined relative to their effect on the intended strategy; costs are defined relative to their opportunity cost, which is the benefit forgone by not using these resources in the best of the available alternative that cannot be undertaken if the resources are used in the project (resp. the operation). The forgone benefits are in turn defined relative to their effect on the intended strategy. By defining costs and benefits in this fashion we try to ensure that acceptance of a project (resp. the operation) implies that no alternative use of the resources consumed would secure a better result from the perspective of the enterprise business intended strategy.

Project (resp. Operation) economic and financial analyses are designed to permit project-by-project (resp. operation-by-operation) decision-making on the appropriate choices between competing uses of resources, with costs and benefits being defined and valued, in principle, so as to measure their impact on the intended strategy of the enterprise business.

Once the (production) line activities (projects or operations) that matter the most have been selected, their numerical relative scores are linked back to key differentiator performance measures; hence to the enterprise business intended strategy demands This is the means by which the enterprise business executives, managers and leaders can measure how well and to which extent the intended strategy has been realized.

8.3.6 Assess Alignment and Implementation

The sixth step of the strategic planning process is concerned with assessment of planning and implementation measures, especially the establishment of a methodology for carefully assessing and reviewing planned activities on an ongoing basis to make sure that what is done each day reflects the intended strategy that the enterprise business has agreed to pursue. The methodology must include systematic identification of progress and problems and a mechanism to initiate corrective actions for deviations from the plan and intended results. In this step, the overall value of the alignment activities (projects and operations), the fit with the intended strategy demands, and the feasibility of the programs are assessed in terms of resources, risk and cost involved on an ongoing basis.

8.3.7 Formulate, Approve and Communicate Alignment Documents

The seventh and final step of the alignment process is concerned the formulation and approval of alignment documents, which form the basis of strategic alignment and control. We have included it here as a part of the alignment process sequence, because in practice it often happens that realization of alignment fails because they are not well documented.

Indeed, aligning the enterprise components through QFD generates a wealth of information. It is important to produce alignment documents which are effective as guidelines for the enterprise business executives, managers and leaders. To fulfill this purpose, the resulting alignment documents must be formulated both concisely and very precisely.

Well documented alignment makes strategic control a simpler matter as well as facilitating strategy review, which will at some time become necessary. It also offers to those not involved in the alignment project the information they need in order to properly understand decisions.

8.3.8 Conclusion

The Quality Function Deployment (QFD) constructs provide an effective and efficient planning methodology to align the components of an enterprise. Performing the planning process well delivers a higher quality result to the enterprise business by merging three important concepts:

1. *Transition from enterprise business intended strategy language to substitute performance measures language and priorities.* The methodology provides a structured transition from the intended strategy demands to the technical specifics with which employees at different organizational levels within the enterprise business can act upon. And, using the enterprise business stated importance rankings of intended strategy demands, it is possible for the enterprise business executives, managers and leaders to organize and allocate their limited resources toward the goal of maximizing their impact on the intended strategy most important wants.
2. *Rational Representations of Relations between the intended strategy and the (production) line activities (i.e., projects and operations).* The methodology uses a representation of the transition that is easy for an individual or a team to relate to and understand because it is both graphical and rationally structured to demonstrate all of the transitions.
3. *Knowledge Gained from a Multifunctional, Interactive Production Line Team.* The more diverse, knowledgeable and interactive the alignment project team is, the more robust the resulting alignment. The proximity of alignment project team members plus their varied backgrounds encourages an integrated approach to the resulting alignment in which no factor is overplayed on the one hand and no key factor is ignored on the other.

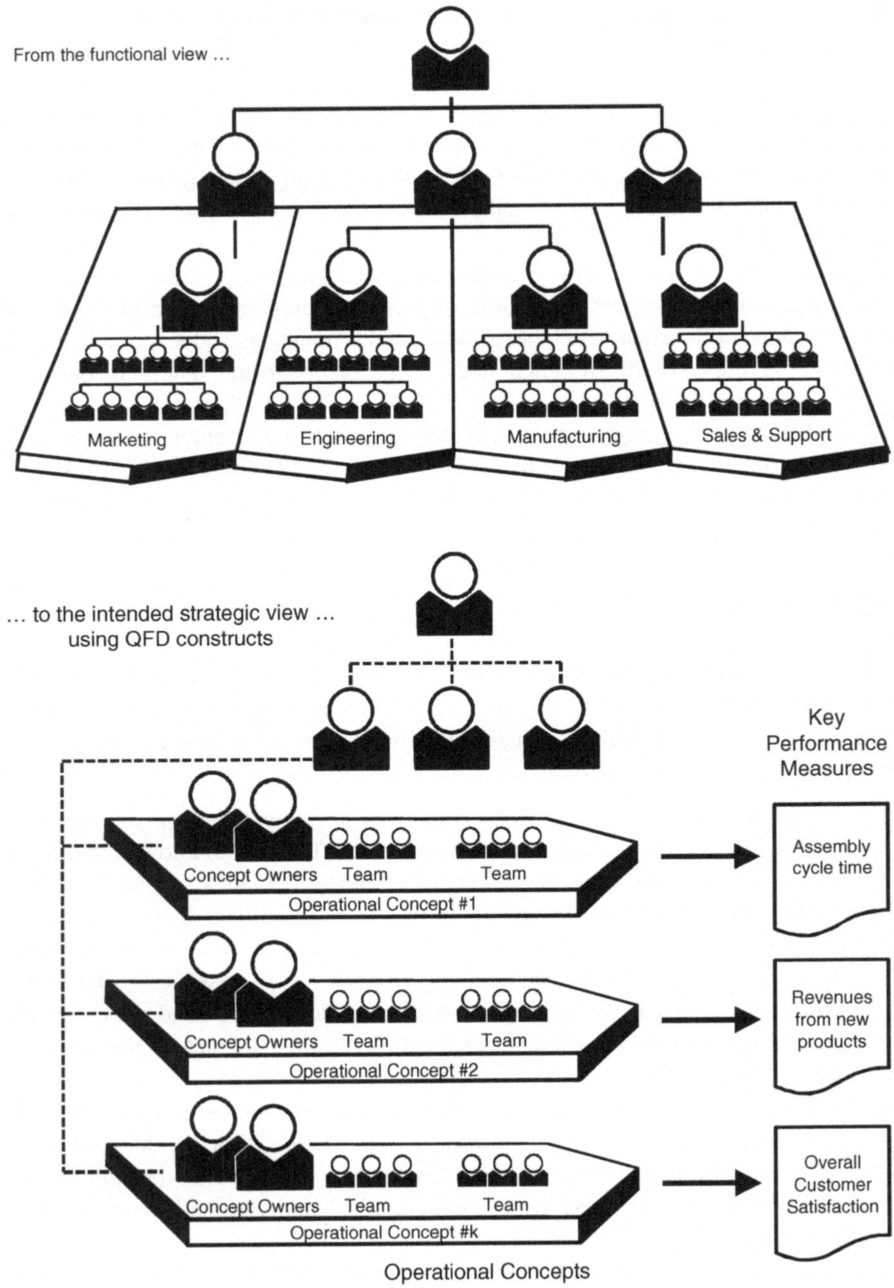

Fig. 8.8 From functional view to intended strategic view

Through the Quality Function Deployment (QFD) constructs, enterprise business executives, managers and leaders understand all phases of the alignment process, including what resources will be needed, to what level and at what point in time. The outcomes resulting from QFD alignment process are much more closely aligned to the needs and desires of the enterprise business.

Some benefits of QFD methodology to enterprise business alignment are as follows:

1. It leads to a better enterprise business alignment than will have been achieved otherwise.
2. It gives this better alignment outcome faster than will other methods.
3. It typically requires fewer resources.
4. It gives definition to the alignment process, helping employees' at all organizational levels to stay focused and effective; giving them greater ability to see and understand how they contribute to the enterprise business intended strategy.
5. It allows for easy management and peer review of alignment activities as they progress, with graphical representation of the different sets of information driving the alignment as well as the relations between information sets.
6. It leaves the enterprise business very positioned should it need to improve upon its results for the intended strategy.
7. It provides a mechanism to view of the enterprise business from its strategic perspective, as illustrated in Fig. 8.8.

Organizational alignment cannot be achieved overnight but it does not have to take a long time either. It just requires a solid plan, focused effort and commitment of employees at every organizational level.

8.4 Commitment

Commitment relates to the human involvement through innate willingness to follow and contribute to achieving alignment plans. It is the employees at every organizational level who will ultimately determine the effectiveness of an enterprise business alignment, because alignment is of no value without human involvement. It is the employees, as human beings, not machines, who act upon the technical specifics resulting from translation of the intended strategy demands. Without people, alignment plans would just be kept in a repository or in a report and not be executed at all. The human involvement brings crucial factors like attitudes, motivation, capabilities, and commitment into implementation of alignment plans.

In a previous chapter, we have indicated that in most enterprise businesses that have not yet reached the "Continuous Improvement" maturity stage, efforts to increase income, lower expenses, and maximize profit in the short term are built around employees "compliance" – the forced adherence to plans created through manipulation, punishment, and coercion. They do not require commitment – the innate willingness of people to follow and contribute. Either people comply with the instructions, or they know they will be at odds with their manager.

Why do enterprise businesses at the "Continuous Improvement" stage of maturity, albeit a minority, succeed in their alignment initiatives and intended strategies while most that have not yet reached the "Continuous Improvement" maturity stage stumble? Because there is "genuine commitment" and then there is "compliance" masqueraded as commitment. How to generate "genuine commitment" on the resulting alignment in an ongoing basis is a challenge facing most enterprises.

Genuine commitment only occurs when employees are willingly and passionately sharing in the ownership and accountability for achieving new levels and standards within an enterprise. It starts with the enterprise business executives, permeates through the managers, and then "infects" everyone with a willingness and urgency to reach new heights through personal improvement. There are two key dimensions to gaining and perpetuating commitment to an enterprise business alignment and its intended strategy: the content and the context.

8.4.1 Alignment Content and Context

The Quality Function Deployment (QFD) methodology, illustrated in the previous sub-section, outwardly set everyone straightforward on the same page about the alignment content, hence the intended strategy content. Getting people behind the alignment does not begin and end with perfecting the alignment content. The alignment content is implemented in a social, economic and environmental context within the enterprise, and has intended or unintended positive and/or negative impacts.

The context of an improvement initiative reflects how the initiative is perceived by employees and therefore how they respond emotionally to it. How employees respond to an improvement initiative is largely a function of how the initiative affects them. Furthermore, the context of an improvement initiative can make the difference between employees being committed and energized by the initiative or employees just minimally complying with it, and even using it for their own personal benefit.

Thus, the alignment project team must also focus on the cultural and social environmental context, which largely determines effectiveness of the alignment plans. Several factors can affect this context, two of which strongly influence it. These are: the climate within the enterprise business and the alignment history.

The climate within the enterprise business: As indicated in a previous chapter, it is the prevailing "atmosphere" within the enterprise business, the social-psychological environment that profoundly influences all behavior, and it is measured by employees' perceptions. Regardless of the improvement initiative been considered for implementation, the prevailing "atmosphere" is what best "defines" an enterprise business to employees. It reflects perceptions on a variety of dimensions, including, among others:

1. The extent of formality (hierarchical structure) versus informality
2. Trust versus distrust (and cynicism) of employees

3. Open versus closed communication
4. Controlling versus collaborative decision making
5. Inward-looking versus outward-looking
6. Past focus versus future focus
7. Task-focus versus people-focus
8. Change versus rigidity
9. Risk-taking versus risk aversion

The alignment history within the enterprise business: relates to the enterprise business experiences with alignment to intended strategies and is much more influential in terms of shaping actions and behaviors than those of alignment content. As Dean Spitzer pointed out in his excellent book on performance measurement: "Both individuals and enterprise businesses are creatures of habit. Our experiences tend to shape what we do subsequently – what we will embrace and what we will avoid – and the same is true for enterprise businesses" (Spitzer, 2007). Throughout an enterprise business, its history tends to send powerful messages regarding expectations for present and future practices.

Clearly enterprise businesses with positive alignment experiences are going to be more open to implementation of the alignment content than those with negative experiences. But it is also a matter of what kinds of experiences have taken place (i.e., routine versus innovative), which will affect how alignment content will be used today. As with any improvement initiative, the major determinant of how eagerly members of an enterprise will embrace the alignment plans is how the "consequences of previous alignments" have personally affected people in the enterprise.

Figure 8.9 illustrates how the alignment initiative content and context shape commitment and compliance within an enterprise. Implementation of an improvement initiative in enterprises within which alignment to intended strategy is well structured and comprehensible (high on content) but within which the social, economic and environmental context has intended or unintended negative impacts on employees (low on context) will at best produce an environment of uninspired compliance.

While employees may well understand the alignment initiative plans and believe in their correctness, they won't believe in their own ability to implement the plans (or implement them without harmful effects). Hence, they will resort to "going along" with little sense of ownership, enthusiasm, and commitment. In the absence of ownership and responsibility, enterprise business executives and managers will resort to managing through dictate, mandate, and command and control; characteristic of enterprises at low stage of maturity. This will undermine people's desire to go the next stage of alignment maturity and to produce exceptional results. This may seem efficient, but ultimately it will be ineffective.

Conversely, if the alignment context is high and the content is low, employees will be highly motivated to make a weak alignment initiative work. Eventually, however, their excitement will not be enough to overcome the bad plans and weak structure. The result will be failure of implementation coupled with cynicism and resignation. When both the alignment content and context are low, employee

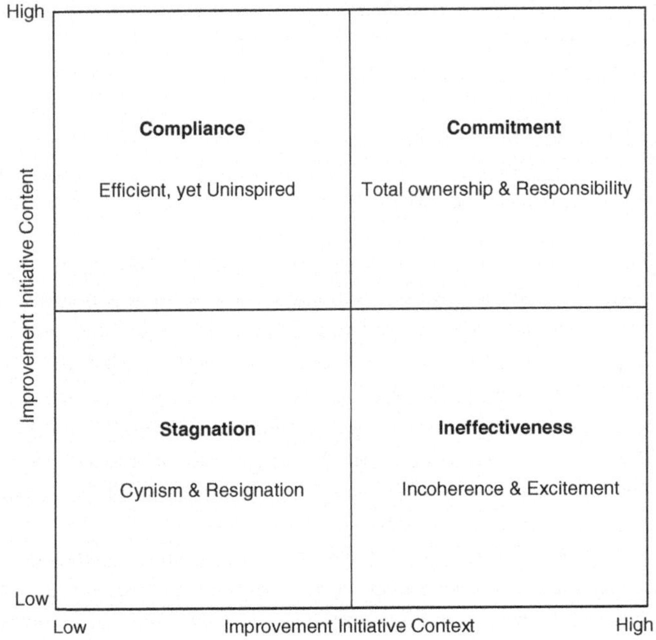

Fig. 8.9 Influence of initiative content and context on commitment

morale will hit the lowest possible level or absolute bottom. Believing in neither the alignment plans nor the enterprise business executives, managers and leaders behind it; employees will be hugely cynical and often oppositional.

But when the alignment content and context are both high, employees' at all organizational levels will believe in the alignment plans and their own ability to act upon the alignment technical specifics to make then happen. This will produce a state of commitment, one in which everyone understands and believes in the alignment to the enterprise intended strategy and feels total ownership and responsibility to make it happen. Thus, a positive context and a well structured and comprehensible content are key dimensions to gaining and perpetuating commitment to an enterprise alignment and its intended strategy.

8.4.2 Assessing Commitment Capabilities

In general, an enterprise business capability is measured by asking the right content and contextual questions. The same is true for assessing the commitment capabilities within an enterprise business. There are proven assessment tools and surveys available in the media to help gauge commitment capabilities within an enterprise business. The Gallup Q12 is a particularly noteworthy tool where a 0.2 improvement along a five-point scale has been statistically proven to correlate with an increase in employee productivity. The survey form in Table 8.1 below, adapted

Table 8.1 Enterprise commitment capability questionnaire

#	Observation	Rating
01	Every employee, regardless of its position in the organizational chart, can cite the enterprise mission and intended strategic goals	
02	Informal conversations are about how best to reach goals, not about confusion and cynicism about direction	
03	There are clear structured and integrated alignment and goals among departments, and not conflicting and self-defeating goals	
04	There are no conflicting priorities, warfare, turf battles, reversals of decisions, and open refusals to cooperate within the enterprise	
05	People never excuse poor results by claiming the goals made no sense	
06	Surveys and feedback never indicate that employees feel uninvolved and are never asked for their opinions or input	
07	Commitment at meetings is matched by behavior and actions; there is no "lip service" agreement followed by inaction	
08	Executives and managers never have to resort to punishment, coercion, or threat to create more support for initiatives and decisions	
09	People at all levels believe the enterprise alignment and key initiatives make sense and are appropriately resourced	
10	People believe management statements and communications, and there is no cynicism about trust and honesty	
11	Employees feel free to ask questions about what they are told and even to challenge things they hear	
12	People feel free to take risks and are not intent on minimizing or avoiding risk by covering oneself from criticism	
13	At the end of the day, real progress and movement can be empirically observed, as opposed to "going through the motions"	
14	Leaders are seen to be in the front, taking risk and showing direction, taking blame but sharing credit	
15	"Whistle blowing" is encouraged, and there are no repercussions for identifying poor performance or bad implementation of technical specifics	
16	Decisions are made, communicated, and implemented with no vacillation, reexamination, or recrimination	
17	There is ready admission of mistakes, wrong direction, and error, and people seek cause, not blame	
18	Difficult and contentious issues are rapidly and effectively raised and addressed	
19	Management follows through on its commitments and proactively communicates when commitments must be modified	
20	People are seen to be promoted based on achievement, not tenure or low profile or political connection	
21	There is lateral communication, so there is no need to stay within hierarchies or silos; "turf" is subordinated to results	
22	Executives, managers and leaders earn the respect of their people and are accessible and visible, and while there may be disagreement, there is always respect	
23	Employees go the "extra mile" for their leaders in terms of workload, hours, and responsibilities, without resentment or complaint	
24	Employees work with intensity for the enterprise good, not personal goals, and do not demand personal reward for every job	

(continued)

Table 8.1 (continued)

#	Observation	Rating
25	Employees believe they are paid competitively, respected for their work, and promoted based on merit	
26	Succession planning and career development plans are in place and are actively monitored by senior management quarterly	
27	There is a mentor program, formal or informal, to help employees deal with challenges and provide private support	
28	Employees feel valued and recognized for the work they do.	
Use the standard five-point rating scale:		
5 = Strongly agree; 4 = Agree; 3 = Neither agree nor disagree; 2 = Disagree; 1 = Strongly disagree		
Interpretation key:		
Highest score is 140		
115–140: Your enterprise business may just be at the cutting edge of alignment and commitment, or you may be kidding yourself about how well your enterprise business is doing. Very few organizations score this high		
98–114: Your enterprise business is doing well, with some room for improvement. Focus on those areas where your enterprise business is weakest for improvement		
70–97: Your enterprise business is in dangerous territory, because it got here by scoring "3" in all areas, which is not very good. You might conclude that your enterprise business is average. In fact, it is probably are shining in one or two contextual areas and doing poorly in the others		
50–69: Your enterprise business has major problems, and it is best to focus on one weak area at a time. We suggest, perhaps counter intuitively, that your enterprise begins with addressing the context of alignment initiative to build credibility, and then focus on improving the alignment content		
Below 50: Scores in this range indicate alignment and commitment weakness: high turnover, destructive warfare, political influence, coercion, etc. These scores are typical of enterprise businesses at low stage of maturity, authoritarian and where employees are viewed as expenses, not assets		

from Josh Leibner (Leibner, Mader, & Weiss, 2009), can be used to assess the current status of commitment within an enterprise business.

8.5 Conclusion

In this chapter, we have provided a closer look at the force that really drive better enterprise business alignment and improved business results. While a clear vision of where the enterprise business is going, an actionable roadmap to get there, and a system of performance measures that provides feedback on progress and identifies opportunities for improvement are all important for success, enterprise alignment is the secret ingredient that is often missing in many enterprise businesses that have not yet reached the "Continuous Improvement" stage of maturity.

One of the most valuable assets of an enterprise business is the synergy that diverse resources and capabilities provide. It is crucial to appreciate how important it is for components of an enterprise to interact. Enterprise businesses at lower stage of maturity can take advantage of much of the synergistic potential at their disposal and use it on an ongoing basis through QFD alignment.

Very few enterprise businesses have established a social or organizational context that is sufficiently supportive of the alignment of their components. Using QFD constructs, alignment of enterprise components within a positive context and strong social aspect will do more than anything else to break down the silos that are keeping enterprises from realizing a "Continuous Improvement" transformational potential.

Team Development and Management 9

Executing the activities resulting from an enterprise business alignment requires a diverse mix of individuals who must be integrated into effective sets: groups and teams (for example: project teams, workgroups). These sets are formed to ease the work pressure on the individual and achieve the alignment activities within the desired time frame. An effective set often outperforms individuals within in an enterprise, because high performance within an enterprise business requires multiple skills, judgments, and experiences. Of course, there are certain tasks at which individuals will always outperform a group or a team; for instance, where talent or experience is the critical performance factor required to achieve an activity. Our purpose in this chapter is to provide guidelines for building a cohesive team.

9.1 Defining a Team

A group within an enterprise business usually comprises a three or more people who recognize themselves as a distinct unit, department, or function, but who actually work independently of each other to achieve their organizational goals. For example, an enterprise business unit may have a client services group, with one person focusing on local clients, one person focusing on regional clients and a third person assisting those individuals.

Within enterprise businesses, groups tend to be permanent fixtures with ongoing goals or responsibilities to sustain businesses. With group work, members have a shared knowledge of the group's objectives, but specific tasks or responsibilities are assigned to different individuals. By separating work into groups – such as one devoted to marketing, one devoted to finance, one devoted to legal and procurement, one devoted to engineering, etc. – individuals within those groups are able to maximize their expertise on a long-term basis.

In today's culture, the word "team" has come to mean many different things. In some fields, a team is an entity merely because it exists, regardless of how well it performs. In business settings, the use of the word "team" implies some level of

A. Van Aartsengel and S. Kurtoglu, *A Guide to Continuous Improvement Transformation*, Management for Professionals, DOI 10.1007/978-3-642-35904-0_9,

exemplary performance. If we are going to invest a significant amount of energy and effort into building a high performance team, we should better be clear about what we are building. We shall define a team as:

> *A group of three or more people committed to a common purpose and working interdependently to produce exceptional and synergistic results for which they hold themselves mutually accountable.*

Working interdependently demands cooperation. Note the prefix "co-" in cooperation and collaborate. Co means "with," "together," or "jointly." The prefix "syn" in synergy also means together. Synergy is the product of cooperation. In an interdependent relationship, the individual members have come to understand that achievement of their personal goals can only be sustained by serving a common purpose, which delivers benefits to themselves and many others. Here, each of the individual members respect and value the uniqueness of the other, as well as appreciating how their individual skills, talents and resources can combine to achieve an overall result that is better than they could achieve by themselves as independent individuals. When inter-dependence exists, the individual members are able to harmonize their efforts so as to achieve synergies (creative cooperation) of the highest order. The results are enriching to each of the individual members and the group sustains itself into the future through positive reinforcing cycles.

The purpose of any team is to accomplish an objective and to do so at exceptional high levels of performance. A team comprises three or more people who may come from different functions or departments within an enterprise business unit, but they collaborate together over time to achieve some set purpose, goal or project. For instance, before a business creates a new product, it may organize a team composed of people from all departments – engineering, manufacturing, finance, legal, marketing, etc. – to consider all aspects of the potential new product to avoid costly surprises down the road. Collaboration between individuals that form the team is not an end in itself but rather a means to an end. Therefore, a team must ultimately be judged by its results.

The primary difference between a team and any other type of group is the synergy circumstances from which the outcome of the team effort results. With a team, individuals recognize the expertise and talents of others needed to achieve the team's goal and they work interdependently and cooperatively. Many groups have a common purpose; most even see some level of cooperation. But in true team, the combination of factors and the intensity and consistency with which they are applied allows a team to experience results on a regular basis.

Enterprise businesses form teams to tackle specific – and usually temporary – goals or projects with the intent of leveraging the collective expertise of a variety of individuals. Because experts from various departments are involved, teams can avoid potential problems early on in projects. For instance, a team of only engineers may create a new product but may not understand whether it's affordable until someone with a finance background completes a "return on investment" or ROI analysis on its feasibility. Having a finance member involved in the team from the beginning will help the engineers to create an affordable product in the first place, saving time and resources. Teams can be very productive because involving individuals with different talents provides teams with increased opportunities to work more efficiently.

As an enterprise business moves towards a "Continuous Improvement" stage of maturity, it uses teams as the load-bearing beams of its organizational structure. They differ in terms of the tasks that they are aiming to accomplish. Within an enterprise business, the tasks of a team can be classified into three categories: production, idea-generation, and problem-solving:

1. Production tasks are those tasks designed to transform tangible inputs (raw materials, semi finished goods, or subassemblies) and intangible inputs (ideas, information, knowledge) into goods or services.
2. Idea-generation tasks are those tasks designed to generate, develop, and communicate new ideas, which are abstract, concrete, or visual, during all stages of a thought cycle: from innovation, to development, to actualization.
3. Problem-solving tasks involve goal-directed thinking and action in situations for which no routine solutions exist. A problem-solving task has a more or less well defined goal, but it is not immediately obvious how to reach it and the incongruence of goals and admissible operators constitutes the problem. The understanding of the problem situation and its step-by-step transformation, based on planning and reasoning, are the essentials of Problem-solving tasks.

As an enterprise business executive, a project manager or team leader, you rarely inherit a fully fledged and effective team at onset of a project or an operation work. More often than not you will likely inherit one that is already misfiring or you will have to start by building your team from scratch. The practical constraints that you will encounter when assembling your team will make this a challenging task. Some of the following might sound familiar to you:

1. Budget constraints preventing much-needed recruiting. Or conversely a generous budget fuelling unrealistic expectations of a fast ramp up.
2. Projects being used as a dumping ground. Other colleague managers using your new team as a convenient home for staff that they are not really sure what to do with.
3. Selfish colleague managers who monopolize the best staff. They hold onto the enterprise business star performers even when their skills and experience are desperately needed elsewhere.

All this is invariably against a setting of an acute sense of urgency to get a team up and running for an improvement intervention. Building and developing the right team – as far as it is realistic – is one of the factors critical to the success of any project.

To maximize the performance of groups and teams within an enterprise business, it is important to understand how they develop and how their dynamics impact the overall performance. Groups and teams both have their own importance and social relevance within enterprises. In a group, people may work independently; taking responsibility for assigned tasks. But real synergy is accomplished when people work interdependently as a team.

9.2 Team Development: The Challenge of Building Teams

A weak, uncooperative team is not just unproductive for the enterprise business; it can make your work a daily grind of frustration as well as resentment. People burn out, blow up, or quit their employment because of negative interpersonal dynamics on teams. Conversely, many people cite the strength of a team or their terrific teammates when recounting how they survived a project when everything seemed to go wrong.

Getting a team to take shape is a difficult task. To the novice it can seem mysterious, the result of good fortune, and certainly unpredictable. But the productivity and joy that come with a high-performance team are too important to rely on good fortune. In reality, every team faces two central challenges, two obstacles to becoming a high-performing team:

1. Teams are formed to accomplish specific tasks, and individual team members must accomplish those tasks together.
2. Teams are temporary and so team members must learn to work together. Not only are teams temporary, but the trend toward teams that cross functional, corporate, and even national boundaries increases the likelihood that a new work requiring a new project team will be made up of people who have not worked together previously.

Understanding these two challenges reveals why some teams work while others never do. Developing trust, respect, effective communication patterns, and the ability to maintain positive relationships despite disagreements takes time. Most important, it takes a conscious effort by the team leader to move a team from a loose collection of individuals to a cohesive unit. Teams that learn to work together to produce effective decisions with efficiency become increasingly bonded and productive throughout the course of accomplishing their specified work.

Several models of group development have been proposed in the literature (Tuckman, 1965; Tuckman & Jensen, 1977; Smith, 2001; Tubbs, 1995; De Dreu & Weingart, 2003a, b; Katzenbach & Smith, 1993; Welins, Byham, & Dixon, 1994; Cannon-Bowers & Salas, 2001; Hackman, 1976; Parker, 1994; McGrath, 1984; Wheelan, Davidson, & Tilin, 2003; Wheelan, 1990; Wheelan, 1994a, b).

The most widely and solidly established such model is Tuckman's "Forming, Storming, Norming, and Performing" model. First published in 1965 and revised in 1970 with the addition of a fifth stage – Adjourning. Tuckman's model explains the necessary and inevitable stages through which a group of individuals must grow before they can function as a cohesive and efficient tasks focused unit. The model has become the basis for subsequent models of group development and team dynamics and management theories frequently used to describe the behavior of existing teams. It has also taken a firm hold in the field of experiential education.

Tuckman's group development model provides a framework for building high-performance teams. High-performance teams are more than merely highly productive. A team composed of experienced, capable people can be very productive until those people hit an obstacle or are confronted with an unexpected challenge. This is the point at which the team either shows its strength or reveals its limitations.

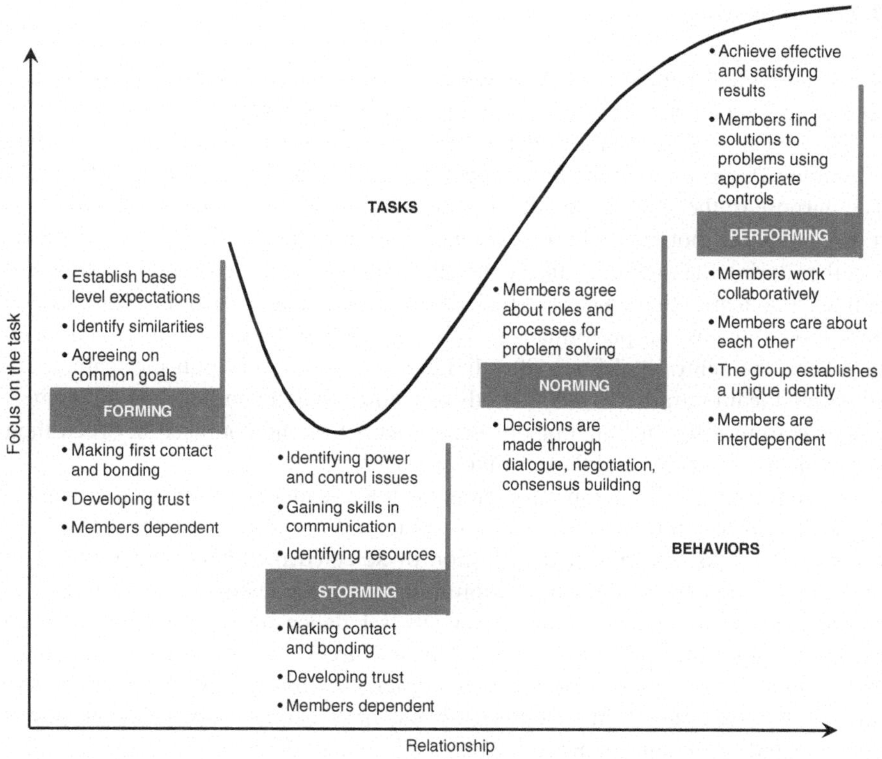

Fig. 9.1 Tuckman's group development model

Tuckman's group development model is best illustrated in Fig. 9.1, which shows the link between group relationships (the horizontal axis) and task focus (the vertical axis). The optimal or "performing" stage is reached when relationships have developed within the group and it has started delivering results with a clear focus on the task.

The model also clearly indicates that it takes time to reach the "performing" stage, and it is normal for groups of individuals to go through ups and downs as they develop relationships, particularly in the early period. As a group of individuals develops maturity and ability, relationships establish, and the leader changes leadership style. Beginning with a directing style, moving through coaching, then participating, finishing delegating and almost detached. At this point the group may produce a successor leader and the previous leader can move on to develop a new group.

Envision the goal of your team as getting from the "Forming" stage to the "Performing" stage. For simple tasks, you could simply grab a handful of people, drop them onto the tasks, and they will somehow get from "Forming" stage to the "Performing" stage without a lot of special attention to team building. That works fine for simple tasks, but if the team has to hold up the weight of a complex undertaking, it must be strong. That is the purpose of Tuckman's group development model – to provide the framework that supports the team for its development.

9.2.1 Forming

When individual people first come together the most fundamental question they look to receive an answer is: "What are we really here for?"

The objective of the team leader at this stage is to establish a clear, compelling "Common Purpose." It is the most important key factor for the success of a team. The purpose of the team is the critical ingredient around which the team will form; it is not only the motivation for its existence, but, like glue, holds the team members together during the inevitable turbulence the team will experience on its journey. It defines the team, the goal of which is to accomplish a specific objective at exceptional levels of performance. Without such a bond of individual team members, the centrifugal force of individual interests would pull the team apart. Ill-defined team purpose sows the seeds of confusion and conflict. Team members assigned from different functions or departments become confused between their departmental priorities and those of the team.

When forming a new temporary team, the team leader is normally interested in the technical and interpersonal skills of potential members that are relevant to the group's tasks, the power distribution of selected members, and whether or not selected members adequately represent relevant constituencies. The key to creating an effective new, temporary team is balance in the attributes of team members, and the presence of needed resources to achieve stated goals. For example, in problem solving and implementation teams, the team leader must make sure that critical enterprise business people with relevant power are selected as team members. Therefore, when decisions are made, non-participating members cannot easily resist. Similarly, the team leader wants to ensure that the required expertise and knowledge exists within the group. This increases the probability of creative problem solving and outcome acceptance by non-members.

At this "Forming" stage, members are positive and polite. Some members are anxious, as they have not yet worked out exactly what work the team will involve. Others are simply excited about the task ahead. The team leader plays a dominant role: he directs and must be prepared to answer quite a lot of questions about the group purpose, objectives and external relationships. Other members' roles and responsibilities are less clear. This stage is usually fairly short, and may only last for the single meeting at which people are introduced to one-another. At this stage there may be discussions about how the team will work, which can be frustrating for some members who simply want to get on with the task.

9.2.2 Storming

Once the "common purpose" has been established, the second fundamental question that individual team members look to receive an answer is: "What is our task?"

The objective of the team leader at this stage is to establish "Clear Roles." High performance teams are also characterized by crystal clear roles. Every team member is clear about his or her particular role, as well as those of the other team members. Roles – functional, specific and general expectations we place on any member of a team if we hope to achieve our objectives at exceptional levels of performance – are all about how we design, divide, and deploy the work of the team. Role issues are invariably one of the top three problems teams face (ineffective processes and communication represent the other two major team problems).

Achieving role clarity is challenged both by old paradigms and new business practices. While the role concept is compellingly logical, many teams find it very challenging to implement in practice. There is often a tendency to take role definition to extremes or not take it far enough. But when they get it right, team members discover that making their combination more effective and leveraging their collective efforts is paramount to achieving synergistic results.

The synergy that team members experience as a team depends in large measure upon three factors (MacMillan, 2001):

1. How the team leader divides the task;
2. How the team leader manages attitudes that can shape how individual members approach their roles and those of others; and
3. How to blend and leverage the different roles on the team against the collective work product.

No amount of team spirit can overcome the wasted motion and energy of poorly designed tasks or survive the frustrations of unclear roles.

This "Storming" stage is characterized by conflict and polarization around interpersonal issues, with concomitant emotional responding in the task sphere. These behaviors serve as resistance to team influence and task requirements, and often results in loss of performance or focus on the task, as the Fig. 8.1 illustrates.

At this stage, the patterns of working start to be defined and some members may feel overwhelmed by how much there is to do, or uncomfortable with the approach been used. Some may react by questioning how worthwhile the goal of the group is, and by resisting taking on tasks. This is the stage when many teams fail, and even those that stick with the task may feel that they are on an emotional roller coaster, as they try to focus on the task in hand without the support of established processes or relationships with their colleagues. The team leader coaches and clarity of purpose increases but plenty of uncertainties persist. The team leader must keep the team focused on its goals to avoid becoming distracted by interpersonal issues. Compromises may be required to enable progress.

9.2.3 Norming

Having established the "common purpose" and clarified roles, the third fundamental question that individual team members look to receive an answer is: "How much autonomy and power do we really have?"

The objective of the team leader at this stage is to establish an "Accepted Hierarchy" and "Accepted Leadership Roles" by providing just the right amount of structure in a likely unstructured environment. Too much and the collaborative spirit will be stifled; too little and the team will flounder and become frustrated.

It is important to stress at this stage that it is often necessary to establish a set of boundaries to the hierarchy. Clear limits need to be established in giving individual team members enough autonomy or encouraging them to assume leadership role. People can slip into irrational behaviors when given freedoms they cannot handle. In this sense, encouraging individual team members to assume leadership roles is not the absence of structure – letting them go off and do whatever they want – but rather a clear structure which enables people to work within established boundaries in an autonomous and creative way.

In addition to the "Clear Roles" established at the "Storming" stage, it is important to establish at this "Norming" stage the necessary ground rules and boundary conditions under which individual team members will be working: what can they decide, what can't they decide? Without the right amount of structure, groups often flounder unproductively, and the members then conclude they are merely wasting their time. The fewer constraints given a team, the more time will be spent defining its structure rather than carrying out its task.

The established hierarchy and leadership roles must be accepted by all individual members of the team, not demanded. As Pat MacMillan reminds us (MacMillan, 2001), being assigned a formal hierarchy or leadership role in a team does not guarantee that the individual members in the team will loyally line up behind you. To be effective, the hierarchy and leadership roles must be accepted by all individual members of the team, and such acceptance is earned, not demanded. When hierarchy and leadership roles are accepted, people are more responsive, more involved, more supportive, and quicker to take initiative.

Gradually, as the group settles into the "Norming" stage, an accepted hierarchy and leadership roles are established. Resistance is overcome in this third stage in which in-group feeling and cohesiveness develop, new standards evolve, and new roles are adopted. In the task realm, intimate, personal opinions are expressed, and the performance of the team gradually improves. At this stage, the team leader facilitates and enables as agreement and consensus are largely formed among members. Critical decisions are made by team agreement. Less critical decisions may be delegated to individuals within the team. Commitment and unity is strong within the team.

At this stage, the team may engage in fun and social activities. It also discusses and develops its processes and working style. Team members come to respect their leader's authority as a leader, and others show leadership in specific areas. They are able to ask each other for help and provide constructive criticism. Team leaders start to view their responsibility as a role rather than a position. They act more as facilitators, networkers, provider of resources, and boundary managers. They are attuned to the needs of the team and serve those needs willingly, knowing that the real "boss" of their team is the task to be accomplished.

The team develops a stronger commitment to the aligned goal, and good progress start to be made towards it. There is often a prolonged overlap between "Storming" and "Norming" behavior: as new tasks come up, the group may lapse back into typical "Storming" stage behavior, but this eventually dies out.

9.2.4 Performing

The objective of the team leader at this stage is to establish "Effective Work Processes" and build "Effective Relationships." Work processes are the "how" the team goes about achieving the "what" in its purpose. They are a sequence of step-by-step actions designed to produce a desired outcome.

9.2.4.1 Establish "Effective Work Processes"

High performance teams are very intentional about both their work and thinking processes. They are clear about what processes they need and then they map and master them. A team can be no more effective than its work processes and the ability of the team to execute them well. Most teams have two basic types of processes – work processes and thinking processes. Work processes (e.g. Work Coordination, Communication, Team cohesion, Decision making, Conflict management, Social relationships, Performance feedback, etc.) are the core processes that accomplish the team's primary mission. For example, sell the product. Thinking processes are process frameworks that facilitate the thinking and discussion of the team as they resolve issues. Too often teams overlook the existence and importance of thinking processes, which need to be addressed with the same degree of deliberateness that the team invests in their work processes.

"Work Processes," like any other dimension of organizational life must be addressed with a determined intentionality. Too often there is too little "design" in work processes within a team. We are often quick to equate any activity with process with little thought about which might be the best activities and what might be the best sequence of step-by-step actions to be performed. In many situations the activities are relatively unconnected with the results we really seek and the result is often a lot of activity but little accomplishment.

It is generally not possible simply to maintain some work processes effective unless preventive measures are set in place. As a consequence of the second law of thermodynamics, we know that a process will tend to erode no matter what, even if a standard is defined, explained to everyone affected, and posted. This is not because of poor discipline by individual team members affected, but due to interaction effects and entropy, which says than any organized process naturally tends to decline to a chaotic state if we leave it alone over time as circumstances change. Interaction effects and entropy continually acts upon all processes to cause deterioration and decay, wear and tear, breakdowns and failures.

When this happens, traditional work processes cease to fit current realities, roles drift out of alignment, relationships become strained, miscommunication occurs, and soon business results suffer.

The pressure to get work done in this no-time, high-urgency world tends to eliminate any thought of taking a few moments to ask, "What processes should we be using? How are we doing and how can we do better?" Team leaders have the power to create time for evaluation by reminding every individual team member that it is okay to call time out to ask those questions.

9.2.4.2 Build "Effective Relationships"

The diversity and differences among the individual team members will most likely preclude close friendships. However, individual team members must be able to withstand the jolts and turbulence of day-to-day interaction, misunderstandings, dropped balls, disagreements, and bad hair days. Building effective relationships by developing mutual trust and understanding of the ways that individual members interact with one another is the cornerstone of teamwork. Effective team relationships provide the climate needed for high levels of cooperation and are characterized by: trust, acceptance, respect, understanding, courtesy, and accountability (MacMillan, 2001).

Trust

As indicated in Chap. 4, we can think of trust as:

> *Expectancy held by an individual or group that promises will be kept and vulnerability will not be exploited.*

It is an "expectation" of dependability and benign intentions typically viewed as a characteristic of personal relationships. Trust is a function of four distinct behavioral characteristics that together form the criteria for its assessment. These are:

1. Being honest (authenticity, forthrightness, veracity, sincerity)
2. Being dependable (reliability, consistency, follow-through)
3. Exercising judgment (ability, capability, capacity, decision making, wisdom), and
4. Generating partnership (Mutual Support, Shared Values and Concerns, Collaboration, Alliance Building).

Trust is the essential quality of building any effective team relationship. Being able to fully trust another (whether or not that trust is ever verbalized) and trust in those processes and values that the team has established in order to achieve its purpose is a function of being genuinely satisfied with each of these four behavioral characteristics. To whatever degree trust is lacking, the source of the gap can always be traced to one or more of these four dimensions.

The development of trust involved being accountable for deviations from the expected performance. So long as those trusted behaved in line with expectations, trust would be reinforced as a result of experience and built progressively over time. It does not need to involve belief in the good character or morality of the individuals and groups involved, merely it needs their conformance to agreed action. It follows our acceptance of an assumed truth about another person or thing.

The development of trust continues and is sustained and enlarged only as future experiences confirm that early perception of expectations to be, in fact, correct. That is, trust builds as experience proves the essential truth of our initial perceptions. Trust diminishes by the reverse; as those trusted do not behaved in line with expectations, we withdraw our trust. Team members will not work interdependently with anyone they do not trust. And without interdependence, there can be no effective division of the task, no leverage of the gifts and skills of individual team members, and, therefore, no synergy. When team members do not trust one another, issues that need to be resolved in meetings become personal, not task oriented. In fact, some team members may not even fully participate due to the fear of conflict. As a result issues are never resolved effectively or efficiently. Without trust, there is no relationship, and there is no team.

In a team relationship, as in any relationship, we trust people because we are comfortable with their character and competence that promises will be kept and vulnerability will not be exploited. By character, we mean our perception of another person's motives, values, honesty, or moral fiber. Competence, on the other hand, refers to the capability, knowledge, and skill of a team member in general and specifically as it impacts his or her assigned role. If we do not trust both the character and competence of a team member, it is unlikely that we will put our desired goals, performance appraisal, compensation, or career into that person's care.

Understanding

Even though team members don't need to know each other all that well personally, they do need to develop a mutual understanding and knowledge of each other's role and potential contribution. Each individual member of the team has his own cognitive understanding and knowledge of the team common purpose and established roles. Team members interact, thereby confirming or sharing their understanding and knowledge of each other's role and potential contribution. This shared mutual understanding and knowledge helps them coordinate their individual knowledge structure within those of the other team members in working through complex problem-solving processes. The deeper the level of understanding and knowledge, the greater the potential for effective collaborative effort, particularly in the less structured, ambiguous circumstances confronted by most management or cross-functional teams. This notion of coordinated shared mutual understanding and knowledge is commonly referred to as a shared mental model (Craik, 1967; Stout, Cannon-Bowers, & Salas, 1996; Brannick, Salas, & Prince, 1997; Stout, Cannon-Bowets, Salas, & Milanovich, 1999; Letsky, 2008; Gettman, 2001; Bergiel, 2006).

Shared mutual understanding and knowledge is a key step to developing trust, and trust is a key element for interdependence and team effort. In the typical cross-functional team, members are generally not performing the same technical task over and over. Team members have no certification as to the competence of other team members. Thus, getting to know and understand each other plays an important role in developing trust.

Shared mutual understanding and knowledge can be expedited by creating opportunities for team members to get to know each another. Because trust is so foundational to any team effort, it is crucial to generally build such opportunities as early as possible. With a little creativity you can find many ways to jump-start members into action.

Acceptance

To accept is to approve of someone, even though that person may be very unlike us. It is easy to accept someone who's like us, but accepting those who are different in values, experiences, manner, and gifts is more difficult. Acceptance is the bridge that connects such differences. It relates to valuing individual differences and is fundamental to the team success. We must accept the fact that there are differences among us in order to motivate our people and build strong teams. Team leaders place a high value on being able to see things through different lenses. Successful problem solving teams and project teams value individual differences as the means of arriving at innovative ways to meeting their goals.

The keys in valuing differences as fundamental to the team success involve two dimensions: internal and external dimensions.

The internal dimension relates to the ability and willingness of qualified individual team members to value differences and share their unique perspectives. This internal dimension requires the following from individual team members:

1. *Recognition of the importance of synergy* – The conviction that the power and strength of the team is greater than the power of any single individual. In business terms, this means that the return on an investment is greater than the contributions made to that investment.
2. *A willingness and eagerness to share knowledge with others and the communication skills to do so* – Willingness implies a positive attitude to other members of the team, a readiness to reply to colleagues kindly. Willingness to share is related to a somewhat passive way of knowledge sharing. Individual team members are willing to contribute to the collective intellectual capital, but they do not have an internal drive to do so.

   Eagerness, on the other hand, implies a positive attitude to actively donating knowledge. We use eagerness to indicate a proactive way of sharing knowledge. People are eager to show what they know, because they themselves consider it valuable and expect their individual performance to be appreciated.

   So for individual team members, who are willing to share their knowledge, the norm of reciprocity is important – they expect others to contribute as well. Therefore, individual team members who are willing to share their knowledge seek to attain a balance between donating and collecting knowledge.

   Eager people, on the other hand, have a strong internal drive to communicate their knowledge, regardless of the team's goals or any directly tangible benefits they can expect from it.
3. *An attitude and belief in plenty* – plenty of opportunity for contribution, plenty of recognition for participation, plenty of reward for accomplishment.
4. *The optimism and conviction that the answer to most team challenges is right around the corner* – Getting around that corner requires looking at things differently.

The external dimension – the overall team dimension – relates to the team's ability to encourage, listen to and use the differences of individual team members to arrive at solutions and processes that far exceed in value and impact that any one member could provide. Conflicts are inevitable in team development. There is no learning without conflict. The ability to focus such conflicts constructively is the highest order of skill in a team. The external team dimension requires three critical elements to maximize the value of individual differences:

1. Clear unambiguous structure focused on the goals of the team. Clear goals, sufficient resources, and effective team composition that recognizes cross functional contribution are all crucial to team success.
2. An investment in team development of communication skills and the development of an understanding and acceptance of different behaviors, values and skills as essential to the best possible outcome.
3. It is crucial that sponsorship of teams be assigned to the top line enterprise business executive who will directly benefit from the team's contribution, as the means of keeping focus on results.

Respect

Respect, or rather the lack of it, seems to be an ongoing concern in many workplaces. Ask anyone in your workplace what treatment they most want at work. They will likely top their list with the desire to be treated with dignity and respect. Lack of respect spells havoc at work. It translates into debilitating costs you may not see on any balance sheet. Apart from fuelling resentments, resignations and absenteeism, it poisons people's experience of work, deprives them of vitality and feelings of self-worth and resilience and drains their purpose and productivity.

Respect is a component that is connected to all characteristics of effective team relationships. Tackling issues of respect therefore has a strong impact. But what does respect actually mean? What do people working in teams mean by respect, and how does respect manifest itself in a team? What drivers or hurdles exist in a team with a tangible influence on respectful teamwork? What methods, tools and strategies can be used to enhance mutual respect within a team and how does one nurtures such an attitude culture (in the long term)?

Respect often seems like sand in the hand – hard to grasp, hard to retain and it all too easily slips through our fingers. Yet ephemeral and slippery as it may be, respect is still critical for constructive relationships, productive teamwork and inspirational leadership. Respect can be perceived as an attitude, a person's ability to see other as equals and acknowledge their values and personal nature. When dealing with colleagues or co-workers, this attitude is reflected in certain situations by the way people observe certain manners. One generally has to consider the fact that people often revert to different behavioral patterns which are dictated by the situation. In today's business environment, people often experience a great deal of stress – with organizational pressures, heavy workloads and vastly differing tasks. How each individual deals with stress, depends on the nature of the person. Consequently when people react to stress, they revert to certain behavioral patterns that they have developed at different stages of the "socialization process."

"Behavioral patterns" is a technique for organizing how one reacts and deals with situations. People revert to a particular pattern of competences they have acquired in the course of their life in order to react appropriately to a given situation. Respectful behavior taps into a variety of personal competences (integrity, self-respect, value-neutrality, authenticity, and helpfulness), social communication skills (ability to communicate, team skills, cooperative skills) and action-based skills (initiative, ability to cope with pressure).

Individual members of a team do not always manage to engage these skills when things become difficult. Conveying respect can be interfered with by organizational and socio-cultural influences. So situations can arise in which it is difficult to communicate one's respect for someone and people are pushed to the limit (e.g. under stress, or when faced with socio-cultural differences).

To respect someone in a team setting means to show honor and esteem for his or her contribution. We must acknowledge that we need each other and we must show equal concern for every member. If one member suffers, we all suffer. If one member is honored, we are all honored. Regardless of whether the team members like one another, they can still treat one another in a respectful manner at work. Maintaining an atmosphere of respect is possible only if the team sets the proper standards and the team follows suit.

Reinforcing respectful behavior brings about several benefits, including improvements in individual team members' loyalty and appeal, satisfaction, ability to change, image enhancements, etc. Improving people's respect for one another at work is not a short-term task. Team leaders should keep in mind the importance of respect in everyday situations and continuously reinforce established fundamental ethical standards and suitable mechanisms to develop respectful behavior. Specific actions can then be planned. Finally, success can be measured with surveys on respect or more informal analytical tools.

The following basic principles serve as guidelines that you can apply to all situations. These principles can help individuals at every level of an organization develop respectful behavior and work more effectively with others to accomplish results:

1. *Focus on the situation issue, or behavior, not on the individual person* – Focusing on the situation, issue, or behavior helps you remain objective when faced with challenges. You can solve problems more effectively, make better decisions, and maintain constructive relationships when you concentrate on the big picture and consider others' points of view with an open mind.
2. *Maintain the self-confidence and self-esteem of others* – Contributing fully is easier in an atmosphere of acceptance and approval. When people feel free to express their ideas without fear of ridicule or personal criticism, they are more willing to take risks and stretch their capabilities. By showing respect and helping others develop their abilities and reach their goals, you multiply your own efforts in the workplace.
3. *Maintain constructive relationships* – The best work comes about when co-workers support one another's efforts. This doesn't mean that you need to be friends with everyone you work with. Your work interactions will go more smoothly, however, if you approach everyone with a positive attitude and communicate support for others. By sharing information, acknowledging problems, and sorting out conflicts, you create strong relationships based on mutual respect. This leads to strong partnerships that will help the team face any challenge that arises.

4. *Take initiative to make things better* – No matter what your role in the team, you can find ways to make things better. By surveying your own area and finding opportunities for improvement, you increase the team's chances for success. You also increase your personal satisfaction by taking control of your work and creating visible improvement. Knowing when to ask for help and when to offer help to others is also key to making things better. Initiative follows naturally when you stay informed and alert to changes and care enough to find solutions to problems.
5. *Lead by example* – As organizations face new challenges, everyone is expected to assume leadership role. Assuming leadership role also means setting a good example – even in the face of setbacks or reversals. Modeling the kind of behavior you want to see in others is the surest way to influence them. By actively honoring your commitments, admitting your mistakes, and staying receptive to new ideas, you will motivate others to do the same.
6. *Think beyond the moment* – For each action or decision, there are consequences. No matter what your role, considering how your actions and decisions will impact others and the team, and avoiding actions that bring personal benefit at the expense of others. When you set compelling goals, make thoughtful plans, and behave ethically, you increase your trustworthiness and dependability. Anticipating the future also helps you prevent minor, manageable problems from turning into organizational crises.

Courtesy

Courtesy is the showing of politeness in one's attitude and behavior toward others and one of the most visible indicators of team relationships. It is a way of acting with people which makes them feel valued, cared for, and respected. Courtesy means to think of how your behavior is affecting others and then do things properly so that they are comfortable. We use courtesy when we are trying to make a good impression.

It is important to show courtesy with individual team members, not just people we are meeting for the first time. We demonstrate courtesy by graciousness, consideration for one another, sincerity, listening, how we talk about teammates who aren't present, and the type of humor we use when jesting with one another.

Practicing courtesy makes every person feel important and acknowledged. No one feels taken advantage of or insulted. The next time they come in contact with you, they want to be around you and help you. Courtesy is like a magnet. It makes you attractive to others. When a person does not practice courtesy, people feel offended and may assume that the person is ill-mannered. They get the impression that the person just does not care about anyone or anything. Rude people are avoided. Others do not feel appreciated. They want to stay away.

Accountability

The final relational quality indispensable in a team setting is "accountability." We shall see accountability as:

A process by which all individual team members agree to be held responsible for the commitments that they have voluntarily made to each other, in addition to individual obligations to their specific roles, and fully accept the natural and logical consequences for the results of their actions.

Accountability is enabled by performance measures. Indeed, accountability is really nothing more than "*measurable commitment*." Performance measures tell you how well you and your team members are performing against commitments – the essence of accountability. As Dean Spitzer indicates (Spitzer, 2007):

...without performance measures, it is difficult to hold yourself – or anyone else – accountable for anything, because there is no way to determine that whatever it is you are supposed to do has actually been accomplished.

Many employees do their best to avoid accountability because it has often been used as ammunition for blame or punishment. It is important to differentiate between "positive accountability" – an opportunity to perform and improve – and "negative accountability" – merely doing what is necessary to get rewards or avoid punishment.

Creating an environment where accountability is clear and fully accepted is a subtle and complex task. The environment of accountability tends to have a major influence on how it is perceived by employees and therefore how they respond emotionally to it. Even if people are not directly held accountable to the natural and logical consequences of the results of their actions, almost everyone feels strongly about accountability. And yet, very few people talk about it in much the same as with performance measures.

The purpose for which accountability is used is the single most powerful determinant of individual team members' reaction to it. Is it being used to provide real understanding, helpful feedback, and to foster learning and improvement – or for justification, reporting, judgment, control, and reward?

"Negative accountability" is when performance measures are used to force performance and punish nonperformance. Because of the flaws and subjectivity in measurement and distrust for those administering it, there is the constant fear that performance measures will be misused. When employees do not feel prepared, are poorly enabled, or view the performance measures as threatening, they will naturally be fearful of the accountability that these performance measures provide. In addition, performance measures will tend to expose those who have traditionally succeeded because of their ability to hide from, manipulate, or finesse the defective measurement systems.

The negative perception of accountability is deeply engrained in most enterprise business cultures. At work, people are typically being help accountable or measured against goals imposed upon them ("These are the targets you are responsible for hitting." "I will be measuring you on or holding you accountable on . . .") and forced into rating categories they feel they do not deserve. Most people in enterprise business teams are accountable for hitting targets. The reaction to hitting targets tends to be quite different from striving to improve one's contribution. Hitting targets leads to a command-and-control orientation and compliance, especially when there are rewards or penalties associated with it.

Managers and team leaders often get very angry, become overwrought or irrational when they see data points that fall below a particular level, and, instead of viewing this as a problem-solving opportunity, they take preemptive action, and sometimes cause heads to roll. Accountability targets (or performance measures) are used, often without being fully understood, to compare teams, or individuals; and employees react by following these performance measures – even if it means going against the common purpose.

In such an environment, everything is focused on hitting the desired target – often by whatever means are available, even if it means bending the established rules. Furthermore, there is a prevailing attitude among many of those accountable for the commitments that they have voluntarily made to each other, in addition to individual obligations to their specific roles, to try to extricate themselves from the responsibility. Because the level of commitment to the common purpose in most teams is very low, it is very tempting to want to delegate it. When this is done, it too often leads to failure to produce exceptional and synergistic results for which individual team members should hold themselves mutually accountable.

Accordingly, for most employees, accountability is viewed, at best, as a "necessary evil." At worst, it is seen as a menacing force that is greeted with about the same enthusiasm as a root canal in dental procedure! When most people think of accountability at work, similarly to performance measures, they tend to think of being watched, being timed, and being appraised. It only takes one snake-bite to make someone fearful of snakes for the rest of their lives – and many people have been bitten more than once by accountability at work. As a result, they tend to see accountability through the lens of negative associations.

In contrast to "negative accountability," there should be no fear of accountability, since it is really just the basis for sound management. It is an agreement to be held to account. It is more than mere commitment and not a judgment. We have established accountability as "measured commitment," because you cannot have accountability without performance measures. This is "positive accountability." Interestingly, the real fear is not of accountability, but of the visibility that measured commitment provides to people who are not accustomed to having to deliver on their commitments.

Accountability, as we have established it, is powerful, and – for better or for worse – what is measured, or held accounted, for tends to get managed. Most employees also seem to intuitively understand that performance measure, hence accountability, is important because their success, their rewards, their budgets, their punishments, and a host of other things ultimately are, directly or indirectly, based on it.

Accountability should actually be a nonjudgmental process. In fact, as soon as judgment is introduced into an environment, there will be almost inevitably some degree of defensiveness occurring. Because of the widespread use of evaluation in enterprise businesses, many people fear that any performance measure can be used against them. Because of previous experiences, people are often suspicions about the motives behind accountability. What does this mean for me? How will they be using the data? In addition, as Dean Spitzer pointed out (Spitzer, 2007):

> *Evaluation has become inexorably linked with demotivating organizational processes and issues such as organizational politics, perceived unfairness, and internal competition for scarce resources.*

It is unfortunate that accountability, as performance measures, is more often viewed as an instrument of control than of empowerment.

Accountability tends to be much more positively embraced by the workforce when it is used as a steering tool, rather than as a grading tool. It will be perceived as a much more positive force – to enable "continuous improvement" transformation, rather than just monitor goal achievement that the target score was attained. If team members are not held accountable for their contributions in such sense, they are more likely to turn their attention to their own needs and advancement rather than collective results.

When effectively practiced, the way that we advocate here above, accountability can be very liberating for both the team and the individual team members. The climate will resound with the understanding that we are all in this together and will succeed or fail as a team. It creates the freedom for team members to proactively share ideas, needs, and to ask for help. In the case of the latter, effective and positive team accountability is like an early warning system that can alert the team if it is getting behind or off course when an individual team member gets stuck, overloaded, or over their head. It makes it easy for team members to yell for help. It also makes it more comfortable for team members to share ideas and suggestions to others outside of their area of expertise or responsibility.

"Positive accountability" does not come easily to people raised in an environment that values the rugged individualist and free spirit. Many view it as a constraint and imposition into their affairs, rather than a contributing element of effective teamwork. As a result, although the concept is often mentioned in team settings, it is seldom defined, and practiced even less.

9.2.4.3 Conclusion

At this "Performing" stage, the interpersonal structure of the team becomes the effective unit of executing task activities. Roles become flexible and functional, and team energy is channeled into the task at hand. Structural issues have been resolved, and structure can now become supportive of task performance. This is the final stage where increased focus on both the task and on team relationships combines to provide synergy. When the team reaches this stage, hard work leads directly to progress towards the shared vision of their goal, supported by the structures and processes that have been set up. Individual team members may join or leave the team without affecting their performance.

At this stage, the group has a shared vision aligned on the enterprise business intended strategy and is able to stand on its own feet with no interference or participation from the leader. There is a focus on over-achieving intended goals, and the group makes most of the decisions with respect to criteria agreed with the leader. Disagreements may also occur but now they are resolved within the team positively and necessary changes to processes and structure are made by the team. The team is able to work towards achieving the enterprise aligned goals, and also to attend to relationship, style and process issues along the way. It functions and operates as a "rowing eights" cohesive unit! At this stage, the team leader is able to delegate much of the work and can concentrate on developing team members as the team is more strategically aware of what it is doing why it is doing.

9.2.5 Adjourning

In 1977 Tuckman and Mary Ann Jensen proposed an update to the popular model, again based on a literature review. They reported that 23 newer articles "tended to support the existence of the four stages" but also suggested a fifth stage. Tuckman and Jensen called this stage adjourning. Adjourning basically involves dissolution; that is, terminating roles, completing tasks, and reducing dependency. It is arguably more of an adjunct to the original four stage model rather than an extension – it views the group from a perspective beyond the purpose of the first four stages.

The Adjourning phase is certainly very relevant to the people in the group and their well-being, but not to the main task of developing a team, which is clearly central to the original four stages. Indeed, teams exist only for a fixed period, and even permanent groups may be disbanded through organizational restructuring. Breaking up a group can be stressful for all concerned and the "Adjourning" stage is important in reaching both group goal and personal conclusions. The break up of the group can be hard for members who like routine or who have developed close working relationships with other group members, particularly if their future roles or even jobs look uncertain.

The value of Tuckman's model is that it helps enterprise business executives and managers understand that groups and teams evolve. It also helps to consider the different problems that may be encountered at different stages of their development. The model also illustrates four main leadership and management styles, which a good leader is able to switch between, depending on the situation (i.e., the group maturity relating to a particular task, project or challenge).

9.3 Realizing Tuckman's Model

In the process of getting your team from the "Forming" stage to the "Performing" stage, critical tasks of the team leader are to:

1. Create an optimal environment for effective development of team members;
2. Develop cooperation or collaborative problem solving capabilities amongst team members;
3. Encourage team members to assume leadership role (as defined in a previous chapter).

9.3.1 Create an Optimal Environment

Relationships among team members, the way meetings are conducted, role and goal clarity – all these factors form the daily environment of the team. People can fight a hostile environment and still make progress toward an established goal, but who wants to? We would rather put our energy into accomplishing tasks than wrangling with uncooperative coworkers or spinning our wheels because the goals and roles are not clear.

The Team leader can influence the daily environment of its team members. He or she can set expectations for conduct during meetings and create opportunities for team members to know and trust each other. In short, the team leader can consciously put into place the attributes that contribute to an optimal environment.

When an optimal environment is established, amazing changes in attitudes, motivation, and performance are not only possible, they are probable. A positive environment promotes trust and respect among team members and increases performance through more productive work habits. Creating this environment requires four specific elements:

1. *Ground rules that describe the work patterns and values of the team.* Ground rules are explicitly stated expectations about personal behavior that reflect the team's values. As Tuckman's model shows, during the "Forming" stage the team wants structure. We meet that team need and begin establishing our work pattern by setting ground rules.
   Ground rules are explicitly stated expectations about team behavior and values. Making these expectations explicit accomplishes three things:
   - Team members understand what is expected of them as a member of an interdependent group.
   - The team has an opportunity to form and own its work pattern;
   - You meet the team's need for structure.

   Ground rules can cover a lot of territory, but generally fall into two categories:
   - *Team values.* Ground rules reinforce specific values by identifying behaviors or attitudes that support the value.
   - *Meeting behavior.* Setting expected meeting behaviors is a classic application of ground rules. Since we often brainstorm solutions, debate alternatives, assign new work among ourselves, and perform other creative work, it is essential that our behaviors demonstrate respect for each other and, at the same time, make productive use of the time we are together.
2. *A team identity built on 'genuine commitment' to a shared goal.* This commitment relies on goal and scope clarity, demonstrated support from the team sponsor, and understanding the strengths and contributions of all team members.
   Forging a team identity is a process that benefits from repetition and attention. As you, as team leader, work to build these elements into your team, recall Tuckman's stages of team development and adjust your style accordingly. The structure of your early attempts to clarify the goals and scope and provide a strong kickoff to the work will be welcomed when the team is "Forming." Likewise, initial attempts to build relationships may be embraced and may seem

productive, yet more attention may be necessary as the team reaches the "Storming" phase. If your investment in team identity pays off, you will find that your team progresses rapidly from "Norming" to "Performing."

3. *The ability to listen.* Problem solving demands an exchange of ideas, which is possible only if team members actually work hard to listen to perspectives that are different from their own.

   When teams form around common challenges and overcome them with creativity and perseverance, strong communication skills are at work. In such a dynamic work environment, no communication skill is more important than listening, because it is through listening that we gain the value of another person's insight. In addition, effective listening builds trust and demonstrates respect during the give-and-take of creativity. As I work to understand your idea, I show you respect that builds trust and increases the likelihood that you will treat my ideas the same way.

   Listening is a personal communication skill. The work of group problem solving requires that every team member have this. Therefore, it is the task of team leaders to model, teach, and coach this skill. Here we highlight some well-known guidelines you and your team can follow. As your team members develop this skill, it will contribute to the overall goal of building an optimal environment.

   Effective listening is a habit that your entire team needs to develop. You can speed this development through several actions early in the work or tasks to be accomplished. The following are guidelines that you, as team leader, can put to use:

   – Pay attention and be attentive of the skill level of the team. That will tell you how quickly and how formally you need to address the listening skills.
   – Plan time to teach the listening skills. Over the course of the work to be accomplished, it is appropriate to spend some time attending to the team's effectiveness. Show a video, pass out and discuss a good article, or bring in a professional trainer to instruct the team.
   – As the team leader, you can demonstrate effective listening, which teaches by example.
   – Look for effective listening behaviors within the team. Point them out as you debrief a team meeting, emphasizing how active listening contributed to a better discussion.
   – Add "active listening" as a desired behavior to your ground rules. Use the ground rules as a reminder during meetings if discussions start to degenerate into arguments.

   Teaching the team to listen pays off rapidly. It is not a difficult skill, but it does take practice and a conscious attempt to improve.

4. *The ability to effectively manage meetings.* Much team work gets accomplished in meetings (or at least it should!). We gather and distribute information, coordinate activities, uncover new problems, assign tasks, and make decisions. Meetings also reinforce team identity, as we gather to make progress on common goals. Productive meetings demonstrate all the characteristics of a high-performance team and produce a result that is beyond what any team members working individually could achieve.

Broken into these four elements, we see that an optimal environment is not merely an abstract feeling; it is a set of observable skills that team leaders can instill. Further, this positive environment produces two important characteristics of the high-performing team:

1. *Personal ownership of the team goal.* Each team member interprets his or her own success in terms of the team's goal. When team success becomes a matter of personal and professional pride, this is a powerful source of motivation and determination.
2. *Strong interpersonal relationships based on trust and respect.* For those of us who have worked on high-performance teams, the friendship during the course of a specified work was far more satisfying than the act of achieving the goal. This element is the most essential and most elusive, because it creates itself. Trust builds trust and respect breeds respect. Trust and respect are essential for people working interdependently because they allow them to rely on one another, which is absolutely necessary if the whole is going to be greater than the sum of the parts.

9.3.2 Develop Collaborative Problem Solving Capabilities

A "problem" is any situation in which the state of affairs varies, or may in the future vary, from the desired state, and where there is no obvious way to reach the desired state. Problems also include situations where nothing has gone wrong yet, but where there is reason to believe that if some action is not taken, something may go wrong in the future. Unless anticipated ahead of time, a "something may go wrong" problem can easily become a "something has gone wrong" problem. Problem solving calls the deployment of strategies calculated to head off foreseeable future problems. It must ultimately eventuate in a decision; that is, a commitment to a course of action that is intended to produce a satisfying state of affairs.

Collaborative problem solving capabilities is the manifestation of teamwork, and the level of collaborative problem solving capabilities drives the level of results. It is important not to see "collaborative problem solving capabilities" as an on-off concept, but a matter of degree. Think of it as a relative concept that allows us to appreciate the dynamic between individual team members and the larger team itself. Team leaders must balance the tensions between the task, the team, and individual members on the team. Too much emphasis on one element at the expense of the others throws the team dynamic out of balance. In many respect, "collaborative problem solving capabilities" is a series of "decision making" patterns that set the pace and direction of the team dynamic.

We have established that teams are established to accomplish specific tasks and that they need to learn to work together to accomplish those tasks successfully. In the process of getting your team from the "Forming" stage to the "Performing" stage, team leaders should endeavor to build this collaborative capability by focusing on four team abilities:

1. *Problem-solving skills tied to an accepted problem-solving process.* A team made up of individuals with diverse skills and styles must agree on the process they will follow for working through problems, both large and small. A commonly accepted problem-solving process enables all team members to flex their styles because each understands and trusts the process.
2. *Understanding and applying multiple decision modes.* Some decisions are made solely by the team leader; other decisions are made by the entire team. These are only two examples of decision modes. Efficient decision making requires that a team understand the possible decision modes and consciously choose which are appropriate for any decision.
3. *Conflict-resolution skills.* Producing superior decisions demands creativity, which necessarily produces disagreement. Mature teams accept and value the inevitability of conflict. They have the skills to leverage conflict to achieve the best decisions while maintaining strong relationships.
4. *Continuous learning.* When innovation and breakthrough solutions are required, the team must embrace and take a certain amount of risk and have the ability to improve its own performance throughout the work been accomplished by learning from both success and failure.

Each of these capabilities can be developed by the team, though not all are simple. Together they create a truly synergistic result: decisions and products that are superior because they are developed by a team with diverse styles and talents.

9.3.2.1 Collaborative Decision Making

We shall think of "decision making" as the cognitive process of making a choice between alternative courses of action (which may also include inaction) in a situation of uncertain events. Decision making involves choosing a particular pathway across the context that lies between the actual and desired states of affairs. It stresses the gathering of information needed to take good decisions.

Making a decision implies that there are alternative courses of action to be considered, and in such a case we want not only to identify as many of these alternatives as possible but to choose the one that is most adequate and effective. Very few decisions are made with absolute certainty because complete knowledge about all the alternatives courses of action is seldom possible. Thus, every decision involves a certain amount of risk.

Our perception of the significance and nature of events leading to decisions after these events have occurred often leads us to make judgments about past choices. During the course of achieving the "common purpose" for which the team was established, individual team members often do not have the benefit of such retrospective information. They usually find themselves navigation in the dark, guided by their individual mental models, incomplete data, and the council of peers and experts.

Chances are that the decisions made by teams are less than adequate and effective. Still, teams can get closer to adequate and effective decision if they have a sound methodology for making decision and some understanding of common decision traps they can avoid. An adequate and effective decision is one that satisfies, to the greatest extent possible, the broadest range of objectives, including constraints, implicated by the identified problem or opportunity for which the decision is called upon.

Decisions are difficult to take when they involve uncertain events, present many alternatives, are complex, and raise interpersonal issues that are difficult to measure but often determine the success or failure of the actions taken. Over the years, scholars have contributed several theories and methodologies to the field of decision making for dealing with these difficulties (Bridge & Dodds, 1975; McGrew & Wilson, 1982; Moody, 1983; Bell, Raiffa, & Tversky, 1988; March, 1994; Adair, 1997, 2009; Teale, 2003; Koehler & Harvey, 2004; Harvard Business School, 2006, 2008; Nutt & Wilson, 2010; Lim, 2010; Brest & Krieger, 2010; Lu, Jain, & Zhang, 2012).

To be an effective member of a team it is critical to understand the team decision making process. The satisfactory team decision process is characterized by a large number of inputs from each individual member upon which other members may build. It is therefore a series of interrelated sub-decisions leading to a final overall decision.

There are essentially two distinct, but complementary, models to decision making. One relies on rationality or reflection, the other on intuition. While intuition is pervasive, reflection is relatively rare because, among other things, it requires considerable cognitive energy. Reflective or rational decision making is informed by intuition at the same time as it corrects for the limitations and biases of pure intuition. Intuition can also be informed by reflection, as happens in the development of expertise. While the processes of intuition are largely opaque to the decision maker, reflection is transparent. For this reason, among others, we begin with a description of the reflective or rational model.

Making Rational Decisions

The rational decision-making model describes a series of steps that team members should consider if their goal is to maximize the quality of their outcomes. There are different types of rational models and the number of steps involved, and even the steps themselves, will differ in different models. An ideal rational model of decision making will consist of the steps or elements described below.

The process is non linear and recursive, beginning with the need to frame the problem in terms of the interests involved and to consider the interests in the context of the particular problem. After completing last step, it would be wise for the team to review the earlier steps, not just because it may have accidentally omitted something, but because the concreteness of positing solutions can reframe objectives and the team conception of the overall problem:

1. *State, or "Frame," the problem or opportunity* – The decision-making process can be triggered by either the observation of a problem or the observation of a unique opportunity that may have presented itself during the course of achieving the "common purpose" and that should be taken advantage of. It is the

observation of problems and opportunities, not their actual symptoms, which gets the decision-making process started. Problems and opportunities may exist all around us, but if they are not perceived and noticed, they do not initiate the decision-making process.

Framing the problem or opportunity for which the decision is called upon sometimes goes about making a wrong decision because the issues are not adequately stated or framed. The statement may mistake symptoms of a problem for the problem itself, or define the problem too narrowly, or define the problem in terms of a ready solution without taking account of the objectives that the team is are actually trying to achieve.

While the causes of some problems are perfectly clear, many others call for analysis or diagnosis. Just as a physician who misdiagnoses the underlying cause of a set of symptoms is likely to prescribe an unhelpful, or even harmful, treatment, so too may a team take useless or counterproductive decisions based on an inadequate analysis of the facts. One way that the team can avoid this is by identifying the problem or opportunity separately from their symptoms.

To clearly frame the real problem or opportunity, an individual team member should be concerned with three basic questions: (1) "What is the problem or opportunity?" (2) "Why should anything be done at all?" (3) "What should or could be happening?"

2. *Establish a context for success* – Every decision is made within a decision context, which refers to the circumstances that form the setting for events, statements, ideas, constraints, preferences, data, social climate, or human factors, and in terms of which it can be fully understood and assessed at the time of the decision. The decision context sets the tone by presenting the purpose of making a decision – achieve a meaningful objective. It reflects how the decision will be perceived by all people affected and therefore how they respond emotionally to it. The right context is critical to making successful choices.
The German philosopher Friedrich Nietzsche is reputed to have said, "To forget one's purpose is the commonest form of stupidity." The best context for a problem is the one that incorporates the broadest possible range of purposes, interests, objectives, and values implicated by the situation. For this reason, the second step in the rational decision model entails a thoroughgoing specification of all relevant interests, conditions, factors, and objectives necessary for success, not just those most readily brought to mind.

In most enterprise businesses, choices are often influences by factors that are antithetical to sound decision making: self-interest, antipathy between individuals, internal rivalries, and alliances based on personal benefits dominate the decision-making process. For example, within enterprise businesses at low maturity stages where command-and-control orientations are predominant, decisions are made in line with the preferences of most powerful individuals. No matter how well informed these powerful individuals may be, every decision is made ad hoc without a consistent approach to handling important choices.

In a positive (decision-friendly) context, that characterizes enterprise businesses at higher maturity stages, robust and rational dialogues take place and the competence of most powerful individuals to take decisions over knowledge workers became less feasible and less legitimate unless these decisions were taken in consultation with the relevant knowledge workers.

3. *Establish and weight decision criteria* – The rational decision-making model has important lessons for teams. First, when making a decision, team members may want to make sure that they establish and weight decision criteria before searching for all alternatives. This would prevent from liking one option too much and setting criteria accordingly.

   For example, let's say that the team has started reviewing proposal for new cars models to be built by an automobile plant factory – ABC Automobile Inc. – before it decided the decision criteria. The team may come across a car model proposal that it thinks really reflects the sense of style shared by all individual team members and make an emotional bond with the proposal of the car model. Then, because of their affection for this car model, the team may say that the fuel economy of the car and the innovative braking system are the most important criteria. After developing and manufacturing it, the team may realize that the car is too small for the target customers friends to ride in the back seat when the front seat is occupied, which was something that the team should have thought about! Setting criteria before searching for alternatives may prevent from making such mistakes.

   Criteria are the measures that we use to arrive at the most adequate and effective decision that best fulfills the purpose. Criteria should not be confused with purpose. The purpose is "what needs to be decided upon and why?" Decision criteria are used to achieve the purpose. The following questions can help the team establish the criteria: "What do we want to achieve in the decision?" "What do we want to preserve?" "What do we want to avoid as problems?"

4. *Generate a range of plausible alternative courses of action* – Once the frame, context and decision criteria have been set, the team should proceed to generates alternative courses of action that might result in goal attainment. This is the stage in the decision-making process that requires the greatest component of creativity and imagination. Ideally, the team should seek to generate as many alternatives as possible and should try to ensure that the set of alternatives is relatively diverse. In this way the team increases the likelihood that some good potential alternatives will not be excluded from further consideration in the decision-making process.

   The team should remember at this step that the best problem or opportunity frame is not necessarily the first to come to mind, and this is true of potential decisions or courses of action as well. Decision making often benefits from a period of divergent thinking about different possible decisions, rather than from rapid convergence on the first seemingly attractive strategic option that sashays by. By generating a large number of alternatives that cover a wide range of possibilities, the team is likely to make a more effective decision in which it does not need to sacrifice one criterion for the sake of another.

There are many methods or procedures that can be used by teams to develop alternatives. Each is designed to improve the decision-making process in some way. Some of the more common group decision-making methods are brainstorming, dialetical inquiry, nominal group technique, and the delphi technique:

- Brainstorming involves team members verbally suggesting ideas or alternative courses of action. The "brainstorming session" is usually relatively unstructured. The situation at hand is described in as much detail as necessary so that team members have a complete understanding of the issue or problem. The team leader or facilitator then solicits ideas from all members of the team. Usually, the team leader or facilitator will record the ideas presented on a flip chart or marker board.

  The "generation of alternatives" stage is clearly differentiated from the next stage, as team members are not allowed to evaluate suggestions until all ideas have been presented. Once the ideas of the team members have been exhausted, the team members then begin the process of evaluating the utility of the different suggestions presented. Brainstorming is a useful means by which to generate alternatives, but does not offer much in the way of process for the evaluation of alternatives or the selection of a proposed course of action.
- Dialetical inquiry is a group decision-making technique that focuses on ensuring full consideration of alternatives. Essentially, it involves dividing the group into opposing sides, which debate the advantages and disadvantages of proposed solutions or decisions. A similar group decision-making method – "devil's advocacy" – requires that one member of the group highlights the potential problems with a proposed decision. Both of these techniques are designed to try and make sure that the group considers all possible ramifications of its decision.
- The nominal group technique is a structured decision making process in which group members are required to compose a comprehensive list of their ideas or proposed alternatives in writing. The group members usually record their ideas privately. Once finished, each group member is asked, in turn, to provide one item from their list until all ideas or alternatives have been publicly recorded on a flip chart or marker board.

  Usually, at this stage of the process verbal exchanges are limited to requests for clarification – no evaluation or criticism of listed ideas is permitted. Once all proposals are listed publicly, the group engages in a discussion of the listed alternatives, which ends in some form of ranking or rating in order of preference.

  As with brainstorming, the prohibition against criticizing proposals as they are presented is designed to overcome individuals' reluctance to share their ideas. Empirical research conducted on group decision making offers some evidence that the nominal group technique succeeds in generating a greater number of decision alternatives that are of relatively high quality.
- The Delphi technique is a group decision-making process that can be used by decision-making groups when the individual members are in different

physical locations. The individuals in the Delphi "group" are usually selected because of the specific knowledge or expertise of the problem they possess.

In the Delphi technique, each group member is asked to independently provide ideas, input, and/or alternative solutions to the decision problem in successive stages. These inputs may be provided in a variety of ways, such as e-mail, fax, or online in a discussion room or electronic bulletin board. After each stage in the process, other group members ask questions and alternatives are ranked or rated in some fashion. After an indefinite number of rounds, the group eventually arrives at a consensus decision on the best course of action.

5. *Evaluate alternatives or predict the consequences of the courses of action and assess their impact on the relevant interests or objectives* – Once a variety of potential courses of action have been generated, the team must proceed to evaluate them rationally. This is done by gathering information regarding each of the alternatives and their likely consequences. More specifically, the team must seek to learn as much as possible regarding the likelihood that each alternative will result in the achievement of the goals and objectives being sought. The team must predict the consequences of each plausible option, and then assess the consequences in light of the objectives for which the decision was called upon.

   Assuming that the problem or opportunity for which the decision was called upon was well defined, evaluation of the adequacy and effectiveness of alternative courses of action should be relatively straightforward. The issue is simply to what extent each alternative alleviates the problem. Using the decision criteria and weights previously identified as important for judging success, the various alternative courses of action can generally be directly compared. However, in addition to simply measuring the end result, the team may also want to consider the consequences each course of action.

   Organizations are made up of real people, with real strengths and weaknesses. A given course of action may require competencies or access to finite resources that simply do not exist in the enterprise business. In addition, there may be political considerations within the enterprise business that influence the desirability of one alternative over another. Therefore, the team may want to consider both the tangible and intangible benefits and costs of each alternative.
6. *Choose and Implement a course of action eventually* – the decision making process comes to a conclusion and a decision must be made. Quite often, this requires making trade-offs among competing interests. The team must select the course of action that optimizes the interests or objectives to be served; that is, make a decision.
7. *Implement, observe, and learn from the outcome of the decision* – Although, strictly speaking, the decision-making process has ended once a decision regarding the most adequate and effective alternative has been reached, it is also true that the decision-making process is no more than a mental exercise if the chosen

course of action is not implemented. Further, issues of implementation are frequently important factors in the choice of an alternative in the previous stages. As implementation of the selected most adequate and effective course of action progresses, it should be monitored, adjusted if necessary, and reviewed to see what can be learned from the experience of its selection and implementation. The decision-making cycle should not end until the team judges the extent to which the chosen alternative has succeeded in addressing the initial problem or opportunity and achieving the goals identified at the outset of the process. If such evaluation indicates success, then the decision-making cycle is concluded. Otherwise, the team must recycle through the decision-making process to generate new alternatives.

The rational decision making process combines elements of divergent and convergent thinking. Divergent thinking expands the range of perspectives, dimensions, and options related to a problem or an opportunity. Convergent thinking eliminates possible alternatives through the application of critical analysis, thereby eventually reducing the number of options that remain open. Divergent thinking conceives; convergent thinking critiques. Divergent thinking envisions; convergent thinking troubleshoots, fine tunes, selects, and implements.

Early in the process, when a problem is being framed, when interests and objectives are being identified, and when alternative solutions are being generated, divergent thinking can bring a great deal of value to the decision making endeavor. Divergent thinking enables us to conceptualize the problem or opportunity from a wide variety of perspectives, so as to permit consideration of the broadest possible array of potential solutions. Divergent thinking helps identify the full range of interests implicated by a particular decision. And divergent thinking inspires innovation in coming up with alternative courses of action. Later in the process, convergent thinking comes into play in analyzing causation, evaluating options, choosing which course of action to implement, and implementing and monitoring the choice.

Making Intuitive Decisions

Most of the time we make decisions without coming close to the conscious, step-by-step analysis of the rational decision making model. In fact, attempting to approach even a small fraction of the problems we encounter in a full, deliberative manner would bring our activities to a screeching halt.

Out of necessity, most of decisions made are intuitive. In contrast with the rational model of decision making, intuitive decisions rely on a process that somehow produces an answer, decision, or idea without the use of a conscious, logically defensible step-by-step process. Intuitive responses are reached with little apparent effort, and typically without conscious awareness. They involve little or no conscious deliberation.

The intuitive decision-making model has emerged as an important decision-making model. It refers to arriving at decisions without conscious reasoning. Nearly 90 % of managers surveyed admitted to using intuition to make decisions at least sometimes, and 59 % said they used intuition often (Burke & Miller, 1999).

When we recognize that team leaders and managers often need to make decisions under challenging circumstances with time pressures, constraints, a great deal of uncertainty, highly visible and high-stakes outcomes, and within changing conditions, it makes sense that they would not have the time to formally work through all the steps of the rational decision-making model.

Yet when team leaders and managers are asked about the critical decisions they make, seldom do they attribute success to luck. To an outside observer, it may seem like they are making guesses as to the course of action to take, but it turns out that they are systematically making decisions using a different model than was earlier suspected. The intuitive decision-making model argues that, in a given situation, experts making decisions scan the environment for cues to recognize patterns (Salas & Klein, 2001). Once a pattern is recognized, they can play a potential course of action through to its outcome based on their prior experience.

Due to training, experience, and knowledge, these decision makers have an idea of how well a given course of action may work. If they run through the mental model and find that the course of action will not work, they alter the course of action and retest it before setting it into action. If it still is not deemed a workable course of action, it is discarded as an option and a new idea is tested until a workable course of action is found. Once a viable course of action is identified, the decision maker puts it into motion. The key point is that only one choice is considered at a time. Novices are not able to make effective decisions this way because they do not have enough prior experience to draw upon.

9.3.3 Encourage Team Members to Assume Leadership Role

While team members focus on the near-term accomplishment of tasks for which the team was established, the team leader must maintain a steady focus on the final outcome of the work been performed and the path toward the goal and the impact on the enterprise business intended strategy. This requires the collective participation of every team member. Every team member must assume leadership role and contribute to the effort.

As we have indicated in a previous chapter, "*genuine commitment*" from any individual is of value only when it is voluntarily and genuinely chosen. We have also established leadership as the accomplishment of a "*common goal*" through the direction of people who are genuinely contributing their creative and productive energies to the process of moving the enterprise business to a higher maturity state.

Commitment of every team member and their involvement should be limited only by his/her analytical and creative capability, and not by his/her position level on the enterprise organizational chart. Leadership as we have defined in a previous chapter, and advocate in this book, is needed not just to make the "Continuous Improvement" transformation contextualized, focused, and interactive through the tasks for which the team was established – and so productive at new levels of effectiveness – but to apply systematically the critical resources needed to realize the rich potentials describes for the transformation of the enterprise business and empowerment of individuals.

9.3.4 Conclusion

In the process of moving the enterprise business from its current stage of maturity toward the "Continuous Improvement" stage of maturity, a key task of the enterprise business executives, its managers and leaders, therefore, is to develop groups and teams following the stages of Tuckman's model. Ironically during our consulting works, we have observed that this route is feared by many managers. However, enterprise businesses at the "Continuous Improvement" stage of maturity place an extremely high value on leaders and managers who can achieve this.

The survey form in Table 9.1 below can be used to assess or uncover common problems that a group or a team might be experiencing.

9.4 Team Management

Team management refers to the comprehensive set of activities followed to establish, implement and improve unity and coordination between the members of a group or a team working towards a common goal – achieve the activities resulting from the enterprise business alignment.

There are a number of different approaches to promote unity and coordination between the members of a group or a team, as well as overseeing or managing their ongoing function. As with many management strategies, there is no one ideal mode of team management that fits every situation and setting. There are few essential characteristics that play a role in any type of team management.

One of those aspects is the ability to accurately identify the strengths and weaknesses that every group/team member brings to the effort. Doing so makes it possible to arrange the essential tasks in a manner that allow people to utilize their skills in areas where they excel, thus moving the entire team closer to the ultimate goal. At the same time, being aware of areas in which different group/team members show some talent or ability makes it possible to cross-train group/team members to handle tasks normally performed by others.

This aspect of team management makes it possible to always have support resources to call upon if a group/team member is incapacitated or unavailable for a period of time. Even if someone is unable to perform assigned duties for a short period of time, tasks are still completed and the group/team continues to move forward.

As an enterprise business executive, a manager or a team leader, your aim is to help your team reach and sustain high performance as soon as possible. If you have opted for Tuckman's model for developing your team, then you will need to change your approach at each stage of the group/team development.

9.4.1 Forming

Direct the group/team and establish objectives clearly. A good way of doing this is to develop a team charter. Team Charters are documents that define the purpose of

Table 9.1 Group/team development questionnaire

#	Observation	Rating
01	My group/team is knowledgeable about the stages of development teams can be expected to go through	
02	Group/team members are provided with a great deal of feedback regarding their performance	
03	Group/team members are encouraged to work for the common good of the enterprise	
04	There are many complaints, and morale is low on my group/team	
05	Group/team members don't understand the decisions that are made, or don't agree with them	
06	People are encouraged to be good group/team members, and build good relationships	
07	Group/team members are provided with development opportunities	
08	Meetings are inefficient and there is a lot of role overlap	
09	Group/team members are encouraged to commit to the enterprise vision, and managers and leaders help them understand how their role fits into the big picture	
10	Group/team members are often given a chance to work on interesting tasks and stretch their knowledge and capabilities	
11	The group/team understands what it needs to accomplish and has the resources needed to be successful	
12	Conflicts, hostilities, or interpersonal issues between members are pervasive issues that do not seem to get better	
13	People feel that good work is not rewarded and they are not sure what is expected of them	
14	Group/team members balance their individual needs for autonomy with the benefits of mutual interdependence	
15	Working relationships across functions or departments is poor, and there is a lack of coordination	
Use the standard five-point rating scale:		
5 = Strongly agree; 4 = Agree; 3 = Neither agree nor disagree; 2 = Disagree; 1 = Strongly disagree		
Interpretation key:		
Highest score is 75		
46–75: You are a solid group/team member working well as part of an effective group/team. Lower scores in this range show that there is room for improvement, though. Read the group/team development stages described above		
31–45: Your effectiveness as a group member/team player and your group/team's effectiveness are inconsistent. You are good at some things, but there is room for improvement elsewhere. Focus on the serious issues below, and you will most likely find that you and your group/team are soon achieving more		
15–30: This is worrying. The good news is that you have got a great opportunity to improve your effectiveness as a group/team member, and the effectiveness of your group/team		

the group/team, how it will work, and what the expected outcomes are. They are "roadmaps" that the group/team and its sponsors create at the beginning of the journey to make sure that all involved are clear about where they are heading, and to give direction when times get tough.

For groups/teams to get off "on the right foot," Team Charters should be drawn up when the group is formed. This helps to make sure that everyone is focused on the right things from the start. However, drawing up a Team Charter can also be useful if a group/team is experiencing difficulties and people need to regain their view of the "big picture." The precise format of Team Charters varies from situation to situation and from group/team to group/team. And while the actual Charter can take on many forms, much of the value of the Charter comes from thinking through and agreeing the various elements.

9.4.2 Storming

Establish process and structure, and work to smooth conflict and build good relationships between members. Generally provide support, especially to those members who are less secure. Remain positive and firm in the face of challenges to your authority, leader's role, or the set goal. Perhaps explain the "Forming, Storming, Norming and Performing" idea so that people understand why conflict is occurring, and understand that things will get better in the future. And consider teaching assertiveness and conflict resolution skills where these are necessary.

9.4.3 Norming

Step back and help the team take responsibility for progress towards the set goal. This is a good time to arrange a social, or a group/team-building event.

9.4.4 Performing

Delegate as far as you sensibly can. Once the group/team has achieved high performance, you should aim to have as "light a touch" as possible. You will now be able to start focusing on other goals and areas of work. Delegating responsibility and authority is one of the fundamental ways to motivate employees and group members. With more responsibility and authority, employees will begin to take ownership. More and more enterprises businesses are giving greater freedom to employees to take initiatives and make decisions.

For example, at one large automobile plant, employees have taken on traditional staff responsibilities for scheduling, quality, safety, hiring, and training. More informally, the final assembly department of the automobile factory tells its employees: "Rule number one is to use your good judgment in all situations; there are no additional rules."

9.4.5 Adjourning

Delegate as far as you sensibly can. Once the group/team has achieved high performance, you should aim to have as "light a touch" as possible. You will now be able to start focusing on other goals and areas of work. As an enterprise business executive, a manager or a team leader, the following steps will help ensure you are doing the right thing at the right time:

1. Identify which stage of the development your group/team is at from the Tuckman's model stages described in the previous section.
2. Consider what needs to be done to move towards the "Performing" stage, and what you can do to help the team do that effectively. Tables 9.2 and 9.3 below help you understand your role at each stage, and think about how to move the group/team forward.
3. Schedule regular reviews of where your group/teams are, and adjust your behavior and leadership approach to suit the stage your group/team has reached.

As an enterprise business executive, a manager or a team leader, you can reduce the difficulties that group and team members experience by understanding what they need to do as they moves through the development stages from Forming to Storming, Norming and, finally, Performing. Think about how much progress you should expect towards the goal and by when, and measure success against that. Remember that the group has to go through the "Forming," "Storming" and "Norming" stages before the team starts "Performing," and that there may not be much progress during this time. Communicating progress against appropriate targets is important if your members are to feel that what they are going through is worth while. Without such targets, they can feel that time has gone by and they have not yet move an inch. Not all teams and situations will behave in this way, however many will – use this approach, but do not try to force situations to fit it. And make sure that people don't use knowledge of the "storming" stage as a license for unprofessional behavior.

9.4.6 Resolving Conflict Rationally and Effectively

Conflict during group development refers to a "perceive divergence of interest, a belief that the parties' current aspirations are incompatible." Because no two individuals have exactly the same expectations and desires, conflict is a natural part of our interactions with others, and conflict is inevitable among individuals in a group development. It is a natural outcome of social interaction that begins when two or more social entities come in contact with one another in attaining their objectives. Relationship among individuals in a group may become incompatible or inconsistent when two or more of them desire a similar resource that is in short supply; when they have partially exclusive behavioral preferences regarding their joint action; or when they have different attitudes, values, beliefs, and skills.

The classical view of conflict within a group development is that conflict is detrimental to group efficiency and therefore should be minimized. This is often

Table 9.2 Summary of group development stages

Stage 1: "Forming"	Stage 2: "Storming"	Stage 3: "Norming"	Stage 4: "Performing"
Individuals are not clear on what they are supposed to do	Roles and responsibilities are articulated	Success occurs	Group members feel very motivated
The mission is not owned by the group	Agendas are displayed	Group has all the resources for handling the task	Individuals defer to group needs
Wondering where we are going	Problems solving does not work well	Appreciation and trust build	No surprises
No trust yet	People want to modify the group's mission	Purpose is well defined	Little waste. Very efficient group operations
High learning	Trying new ideas	Feedback is high, well received, and objective	Group members have objective outlook
No group history; unfamiliar with group members	Splinter groups form	Group confidence is high	Individuals take pleasure in the success of the group – big wins
Norms of the group are not established	People set boundaries	Leader reinforces group behavior	"We" versus "I" orientation
People check one another out	Anxiety abounds	Members self-reinforce group norms	High pride in the group
People are not committed to the group	People push for position and power	Hidden agendas become open	High openness and support
	Competition is high	Group is creative	High empathy
	Cliques drive the group	More individual motivation	High trust in everyone
	Little team spirit	Group gains commitment from all members on direction and goals	Superior group performance
	Lots of personal attacks		OK to risk confrontation
	Level of participation is at its highest or at its lowest		

done by prescribing group structures – rules and procedures, hierarchy, channel of command, and so on – so that group members would be unlikely to engage in conflict. This approach to resolving conflicts within groups is based on the assumption that harmony, cooperation, and the absence of conflict are appropriate for achieving group effectiveness.

However, conflict can also be seen as an instrument of social change and influence rather than a symptom of a breakdown in social relationships within a group development. This is the modern view of conflict within enterprises at the "Continuous Improvement" stage of maturity.

Table 9.3 Leadership action steps between group development stages

Action steps: "Forming" to "Storming"	Action steps: "Storming" to "Norming"	Action steps: "Norming" to "Performing"
Set a mission	Leader should actively support and reinforce group behavior, facilitate the group for wins, create positive environment	Maintain traditions
Set goals	Leader must ask for and expect results	Praise and flatter each other
Establish roles	Recognize, publicize group wins	Self-evaluate without a fuss
Recognize need to move out of "forming" stage	Agree on individuals' roles and responsibilities	Share leadership role in group based on who does what the best
Leader must be directive	Buy into objectives and activities	Share rewards and successes
Figure ways to build trust	Listen to each other	Communicate all the time
Define a reward structure	Set and take group time together	Share responsibility
Take risks	Everyone works actively to set a supportive environment	Delegate freely within the group
Bring group together periodically to work on common tasks	Have the vision: "We can succeed!"	Commit time to the group
Assert power	Request and accept feedback	Keep raising the bar – new, higher goals
Decide once and for all to be on the group	Build trust by honoring commitments	Be selective of new group members; train to maintain a positive context spirit

In this approach, conflict behaviors must occur from time to time in order to demonstrate the will and capacity of action. Conflict itself, especially when innovative alternatives are being analyzed and challenged, is a necessary ingredient in the creative process. Differences among group members are often the catalysts to vigorous debate and creative thinking.

A critical challenge for leaders and their group/team members is how to get the best from the inevitable differences and disagreements that arise during group/team development while minimizing the harm and discomfort routinely associated with conflict. Conflict during group/team development is considered legitimate, inevitable, and an indicator of effective group/team management within enterprises at the "Continuous Improvement" stage of maturity.

Within certain limits, conflict is essential to the group productivity. It can be functional to the extent to which it results in the formulation and creative solution to the right problems or the effective attainment of enterprise intended objectives that otherwise would not have been possible. Little or no conflict in a group may lead to stagnation, poor decisions, and ineffectiveness. On the other hand, conflict left uncontrolled may have dysfunctional outcomes.

Therefore, the more critical issue within enterprises at the "Continuous Improvement" stage of maturity is whether groups/teams are experiencing enough conflict. Too little conflict may encourage stagnancy, mediocrity, and groupthink, but too much conflict may lead to group disintegration. A moderate amount of conflict handled in a constructive manner, is essential for attaining and maintaining an optimum level of group effectiveness. The functional and dysfunctional outcomes of conflict in a group are as follows:

Functional Outcomes:

1. Conflict may stimulate innovation, creativity, mutual understanding, and change.
2. Organizational decision making process may be improved.
3. Alternative solutions to a problem may be found.
4. Conflict may lead to synergistic solutions to common problems.
5. Individual and group performance may be enhanced.
6. Improved self-knowledge.
7. Individual and groups may be forced to search for new approaches.
8. Individuals and group may be required to articulate and clarify their positions.

Dysfunctional Outcomes:

9. Conflict may cause job stress, burnout, and dissatisfaction.
10. Communication between individuals and groups may be reduced.
11. A climate of distrust and suspicion can be developed.
12. Relationships may be damaged.
13. Job performance may be reduced.
14. Resistance to change can increase.
15. Organizational commitment and loyalty may be affected.

There are two types of conflict that may appear during a group development: task and relationship conflicts.

A task conflict (also referred to as cognitive conflict[1]*)*, emerges when group members have differences of opinion but are able to stay focused on solving the problems caused by their differences. Their discussion of issues typically results in higher levels of creative thinking and better decision-making because the issues are more fully vetted (Chen, 2006). A task conflict often occurs during the early stage of decision making and it stimulates creativity.

A relationship (also referred to as affective conflict), occurs when group members spend more time trying to assign blame than on figuring out how to solve problems. It is associated with poorer group productivity and lowered morale (De Dreu & Weingart, 2003). Even when group members try to effectively debate issues it is easy for their efforts to devolve into relationship conflict. Critiques of

[1] Cognitive conflict here refers to conflict regarding the way a group approaches and attempts to solve problems encountered in the course of its development. Viewed most broadly, cognitive conflict is a conflict in way of seeing and thinking a perspective.

ideas can easily be perceived as personal attacks. When this happens, a task conflict can quickly morph into a relationship conflict, with undesirable results.

So the key question for an enterprise executive, a manager or a team leader becomes: "how can group members deal with their inevitable differences in ways which foster constructive forms of conflict while avoiding or lessening the emergence of destructive relationship conflict?"

Thus, conflict in a group has both positive and negative consequences. If the group is to benefit from conflict, the negative effects of conflict must be reduced, and the positive effects must be enhanced. If a conflict is not controlled and handled effectively, the results can be damaging. Conflicting goals can quickly turn into personal dislike and work breaks down. Talent is wasted as people disengage from their work. And it is easy to end up in a vicious downward spiral of negativity and recrimination.

9.4.6.1 Setting Norms to Manage Conflict Effectively

As an enterprise business executive, a manager or a team leader, if you are to keep your group or team working and resolving conflicts effectively through the group development stages, you need to create/reinforce a positive context within your group/team. Group members need to be able to discuss issues openly and candidly. They need to have a sense of mutual responsibility for resolving their problems. To achieve this, you must endeavor to develop the right climate to foster openness and collaboration. This requires the use of constructive communication skills and techniques to keep discussions heading in the right direction. Groups frequently focus on their substantive tasks without taking the time to address process concerns. If you want to keep conflicts from derailing efforts, an essential first step is to establish norms and processes for dealing with the inevitable conflicts your group/team will face.

What kinds of norms do you need to develop/reinforce in order to manage conflict effectively? Several elements are essential in establishing norms for addressing conflict. These include trust and safety, collaboration, and emotional intelligence. At the outset, you must address with the group members how the group will promote and support each of these elements.

Trust and safety – In order to feel comfortable enough to share thoughts and feelings openly and honestly, members of your group must trust each others. Trust develops when members make themselves vulnerable by being honest, open and willing to exchange fresh ideas. Group leaders can support the process of building trust by showing vulnerability themselves and ensuring that the group develops norms for interacting while under stress. A group can enhance the development of trust and safety through a structured disclosure, which enables members to share interests, insights, and experiences safely. In addition, we encourage teams to identify or predict potential "hot topics" to eliminate surprises.

Collaboration – It refers to behavioral integration of individual members. It occurs when members share information freely, make decisions together, and are recognized and rewarded collectively, group cohesiveness increases. When collaboration is practiced consistently, trust is reinforced, and members can debate issues more effectively. One technique we advocate is "preliminary perspective taking" during which members quickly and concisely state their starting views without interruption.

Perspective taking refers to our ability to relate to others. It is our ability to perceive someone else's thoughts, feelings, and motivations. In other words it refers to our ability to empathize with someone else and see things from their perspective. This is a powerful and effective tool for engaging conflict constructively. Most conflicts become more volatile and intense with the failure to acknowledge differences constructively. Simply put, when done well, perspective taking demonstrates a willingness to consider the views, positions and feelings of others. We also recommend periodic team training sessions to practice devil's advocacy, reframing, and brainstorming to build collaborative skills.

Emotional intelligence – Conflict by its very nature often ignites emotions. Negative emotions can easily spread among team members through a process called emotional contagion (Goleman, 2007). When team members are upset, handling conflict becomes very complicated. As defensiveness rises and openness wanes, if the negative emotions are not addressed effectively, destructive behaviors and dysfunctional outcomes soon follow. Individual members can improve their emotional intelligence by utilizing assessment tools that raise self awareness (Hughes & Terrell, 2008; Straw & Cerier, 2002).

For many people, conflict seems to spiral out of control in an instant. The speed and force with which conflict arises can be alarming. Taking the time to consider the type of situations that "set you off" is a great way to improve your readiness for conflict. When you are more mindful of your typical reactions, it's easier to recognize your emotions earlier during conflict. Such awareness enables you to restrain yourself. With self assessment results at hand, establishment of norms that address healthy emotional intelligence will provide a stable base for managing conflict.

9.4.6.2 Thomas-Kilmann Conflict Mode Instrument

Creating/Reinforcing and maintaining a positive context within a group cannot be accomplished unless group members choose to communicate in constructive ways. All too often when confronted with conflict, members behave or respond in destructive ways. Often destructive behaviors take the form of fight or flight responses. Some of the most common destructive types of responses include:

1. *Winning at all costs* – attempting to get "your way" no matter what.
2. *Avoiding* – withdrawing from the conflict and your conflict partner.
3. *Demeaning others* – devaluing others or using sarcastic language.
4. *Retaliating* – actively or passively trying to "get even."
5. *Yielding* – giving in to your conflict partner.
6. *Hiding emotions* – concealing one's true feelings.

Instead of engaging in destructive behaviors, group members must choose to respond in more constructive ways. Admittedly, in the heat of the moment, this may not be an easy task. As an enterprise business executive, a manager or a team leader, you should be able to measure people's behavior in conflict situations. This can be done using Thomas-Kilmann Conflict Mode Instrument.

Thomas-Kilmann Conflict Mode Instrument describes an individual's behavior along two basic dimensions: (1) assertiveness, the extent to which the person attempts to satisfy his own concerns, and (2) cooperativeness, the extent to which the person attempts to satisfy the other person's concerns. These two basic dimensions of behavior define five different modes for responding to conflict situations:

1. *Competing*: Competing is assertive and uncooperative – an individual pursues his own concerns at the expense of others. This is a power-oriented mode in which the individual uses whatever power seems appropriate to win his own position – his ability to argue, his rank, or economic sanctions. The individual stands up for his rights, defending a position which he/she believes is correct, or simply trying to win.
2. *Accommodating*: Accommodating is unassertive and cooperative – the complete opposite of competing. When accommodating, the individual neglects his own concerns to satisfy the concerns of the other person; there is an element of self-sacrifice in this mode. Accommodating might take the form of selfless generosity or charity, obeying another person's order when the individual would prefer not to, or yielding to another's point of view.
3. *Avoiding*: Avoiding is unassertive and uncooperative – the individual neither pursues his own concerns nor those of the other individuals. Thus he does not deal with the conflict. Avoiding might take the form of diplomatically sidestepping an issue, postponing an issue until a better time or simply withdrawing from a threatening situation.
4. *Collaborating*: Collaborating is both assertive and cooperative – the complete opposite of avoiding. Collaborating involves an attempt to work with others to find some solution that fully satisfies their concerns. It means digging into an issue to pinpoint the underlying needs and wants of the two individuals. Collaborating between two persons might take the form of exploring a disagreement to learn from each other's insights or trying to find a creative solution to an interpersonal problem.
5. *Compromising*: Compromising is moderate in both assertiveness and cooperativeness. The objective is to find some expedient, mutually acceptable solution that partially satisfies both parties. It falls intermediate between competing and accommodating. Compromising gives up more than competing but less than accommodating. Likewise, it addresses an issue more directly than avoiding, but does not explore it in as much depth as collaborating. In some situations, compromising might mean splitting the difference between the two positions, exchanging concessions, or seeking a quick middle-ground solution.

Each individual in a group is capable of using all five conflict-handling modes. No individual in a group can be characterized as having a single style of dealing with conflict. But certain people use some modes better than others and, therefore, tend to rely on those modes more heavily than others. Once you, as an enterprise executive, a manager or a team leader, understand these different styles, you can use them to devise the most appropriate approach (or mixture of approaches) for the situation in which your group is in.

9.4.6.3 Conclusion

Conflict management is not just for team leaders; it is a team skill. If your team breaks down when facing disagreement, then it is time to focus on building this skill. This is one topic that the team leader probably should not try to teach to the team. Unless you have been trained in this area, avoid the do-it-yourself approach. Most large enterprise businesses have human resources or organizational development professionals who are educated in conflict management. Invest in this skill early in the team development to reduce "Storming" and speed the transition from "Norming" to "Performing."

9.5 Conclusion

This chapter has emphasized that team members are challenged to work interdependently to accomplish specific tasks within an enterprise business. These tasks can be simple or complex, but every task is, essentially, a series of decisions to make and problems to solve. The greater a team needs to work interdependently, the more its members need to trust each other and the more they need the skills to work together. Since teams are temporary, team members must learn to work together.

The Tuckman's group development framework identifies the factors that must be present for a team to reach its synergistic potential: to deliver more as a team than the individuals working alone ever could. A team leader helps the team establish an optimal environment for daily interaction by setting ground rules for team behaviors, ensuring the team uses good listening skills, practicing good meeting management, and building team identity. To enable the team to face problem after problem together, the team leader should work to improve its collaborative capability, including teaching the team problem solving and conflict-resolution skills.

Process Improvement and Management 10

The purpose of this chapter on "Process Improvement and Management" is not about performing the systematic methodology for process improvement; it is about creating an optimal environment for effective implementation of process improvement and management within an enterprise business.

In a previous chapter, we have indicated that the (production) line activities designed to support realization of the operational concepts include projects and operations that matter the most. Projects and operations have fundamentally different objectives.

A project is a sequence of unique, complex, and connected activities having one goal or purpose and that must be completed by a specific time, within budget, and according to specification. It is a temporary effort undertaken to create a unique product, service, or result. The purpose of a project is to attain its objectives and then terminate. Projects are therefore utilized as a mean of achieving an enterprise business intended strategy. They conclude when their specific objectives have been attained. Operations are ongoing and repetitive efforts, the purposes of which are to sustain the enterprise business. When their objectives have been attained, operations adopt a new set of objectives and the work continues.

Although projects and operations sometimes overlap, both share the following characteristics: they are constrained by limited resources; they are selected following analyses of their added value in terms of costs and benefits to the enterprise business; they are performed by people; and they are planned, executed, and controlled.

10.1 Characterizing and Defining a Process

Another key characteristic that projects and operations also share is that they often use common series of sets of logically related discrete elements (tasks, actions, or steps) with well defined interfaces in order to achieve their objectives. These sets of logically related discrete elements (tasks, actions, or steps) are not goals in

A. Van Aartsengel and S. Kurtoglu, *A Guide to Continuous Improvement Transformation*, Management for Professionals, DOI 10.1007/978-3-642-35904-0_10,

themselves within an enterprise business; they are mean to achieve operations and projects work. We define a process as:

> *A set of logically related discrete elements (tasks, actions, or steps) taken in order to achieve a particular end.*

In this definition, a discrete element, the performance of which is measurable, is meant to be the smallest identifiable and essential piece of activity that serves both as a unit of work and as a means of differentiating between the various aspects of a project or an operation work. Each discrete element is designed to create unique outcomes by ensuring proper control, acting on and adding value to the resources that support the work being completed.

From the perspective of this definition, a process acts on and adds value to the resources that support the activities being completed by a project or an operation work. Furthermore, each discrete element of a process has two aspects:

1. Its operational definition or specific technical content, which is addressed in our next book, and
2. Its context, which is represented by everything else that surrounds and affect the specific technical content.

A process is a set of logically related discrete element (tasks, actions, or steps) taken in order to achieve a particular end. But when most people think of process at work, it is much more than the operational definition or specific technical content of its discrete elements that they are reacting to: it is the patterns of interaction ensuing from the resulting specific technical content, plus the resulting context.

Thus a process is characterized by the patterns of interaction, coordination, communication, and decision making employees use to transform resources into products and services of greater worth. Processes include not just manufacturing processes, but those by which product development, procurement, market research, budgeting, employee development and compensation, and resource allocation are accomplished. Some processes are formal, in the sense that they are explicitly defined and documented. Others are informal: they are routines or ways of working that evolve over time. The former tend to be more visible, the latter less visible.

Processes are defined or evolve de facto to address specific tasks. This means that when employees use a process to execute the tasks for which it was designed, it is likely to perform efficiently. But when the same seemingly efficient process is used to tackle a very different task, it is likely to prove slow, bureaucratic, and inefficient. In contrast to the flexibility of resources, processes are inherently inflexible. In other words, a process that defines a capability in executing a certain task concurrently defines disabilities in executing other tasks. One of the dilemmas of management is that processes, by their very nature, are set up so that employees perform tasks in a consistent way, time after time. They are meant not to change or, if they must change, to change through tightly controlled procedures in order to avoid unproductive habits.

The best way to determine an enterprise business processes from the outside is to imagine the kinds of problems and challenges the enterprise business must have repeatedly faced and solved that have led to its success and defined its daily life.

Telephone enterprise businesses have to build and maintain large, complicated telecommunication networks that must work just about all the time. Automobile car manufacturers have to coordinate a complicated network of suppliers. Medical enterprise businesses have to gain approval for new medical devices etc. For these enterprise businesses to be successful, they must have developed was to face challenges and solve these problems again and again. They need processes that facilitate their ability to get done what has to get done.

There are also less visible background processes that support critical decisions such as where to invest resources, how market research is habitually performed, how financial projections are created, how plans and budgets figures are negotiated internally, etc. Many of these important processes that define an enterprise business' strength are not readily observable to outsiders or, as a matter of fact, to insiders.

Thus, a listing of the recurrent problems or tasks that an enterprise business has successfully, repeatedly addressed is a visible and reasonably accurate proxy for a listing of its processes. This means that if an enterprise business has never confronted a particular problem or task before, an optimized process to complete that task would not exist.

10.2 Importance of Business Processes

Most enterprise business executives struggle with the concept of why business processes are important to an enterprise business. Historically there has been little formal tertiary management education on the opportunities that business processes bring to an enterprise business or the impact on an enterprise business if they are sub-optimal.

Some of the recent literature in the process world has suggested that business processes are so important that the enterprise business structure should be turned upside down to be a process-centric organization, rather than functionally based. It is argued that changing from the traditional functional, hierarchical orientation to a process-centric orientation will mean that enterprise businesses will function with greater efficiency and effectiveness, to the benefit of management, staff, customers and all other stakeholders.

After all, a functional organizational structured view creates a silo effect within an enterprise business, and this often leads to selfish or self-centered behavior by the management and staff of each silo, sometimes to the detriment of other silo's and the enterprise business as a whole. In most enterprise businesses, there is significant effort expended attempting to minimize, or eliminate, this silo effect but it can take years and years to orientate all the management to a more holistic approach and behavior. If successful, the challenge then is to maintain this new found focus as the management and staffs come and go from the enterprise business. If this is not successfully passed from one manager to another, then the enterprise business can regress back again to a silo-like situation. After all, this is how business has successfully functioned for decades.

While a process-centric structured enterprise business can in certain circumstances significantly benefit an enterprise business this not always true. Even if an enterprise business achieved the perfect organizational structure, this is still not a guaranty for its future success. Enterprise businesses are complex and intertwined organisms with no one aspect being dominant or the warranty to solve all its challenges and issues. The continual and sustainable success of an enterprise business is a complex set of interacting events and criteria and much has been written on how to achieve synergy.

Results are driven by the synergy of the eight overarching determining factor of strategic management outlined throughout this chapter. Business processes provide an enterprise business' ability to deliver products and services to customers. In much the same as performance measures, they are the link between all aspects of an enterprise business. Processes are the link between an enterprise business and its:

1. Suppliers
2. Partners
3. Distribution channels
4. Products and services
5. People (personnel)
6. Other stakeholders.

Of course, while a "performance measure" is a necessary condition for success, it alone is not sufficient for it. We still must take informed action. This is performed through a business process. Therefore we see business processes as the central core from which business is conducted, so long as they are supported by the performance measures and resources within the enterprise business.

10.3 Realizing "Process Improvement and Management" Transformation

Enterprise businesses create value as employees use processes to transform inputs of resources into products and services of greater worth. The principle that the resources that support the work being completed, the quality, and the execution time associated with discrete elements of a process can be optimized methodically is the basis of "Process Improvement" philosophy.

And the principle that a methodic management of resulting processes could design the best rational way of performing any activity within enterprises, which would lead to enhanced productivity and profitability, is the basis of "Process Management"[1] philosophy.

[1] The term "Process Management" is sometimes used in the media to describe an organizational or managerial approach (referred to as "Management by Process) to the management of processes and some ongoing operations, which can be reduced to processes. An organization that adopts the "Management by Process" approach defines its activities as processes in a way that is consistent with the definition of process provided above.

Thus, "Process Improvement & Management" refers to the comprehensive set of activities followed to establish, implement and optimize the performance of enterprise processes. It includes defining expectations and accountabilities, setting process capabilities, process performance standards and performance measures, and assessing results. It is the centralized and coordinated management of processes to: obtain the benefits and control not available from managing them individually and, achieve the objectives of operations and projects work necessary to realize an enterprise intended strategy.

Although a methodic optimization of a process can be done separately from a methodic management of processes, process improvement cannot be dissociated from process management within an enterprise. In other words, process improvement alone is not an end in itself. There are two aspects to the improvement and management of processes:

1. The management of processes as an integral part of the enterprise business management, and
2. The management of process improvement.

The management of processes as an integral part of the enterprise business management – As an integral part of the enterprise business management, the management of processes is concerned with achieving the objectives of projects and operations work critical to realizing the enterprise intended strategy. It is performed by line managers owning these processes, while middle managers perform the management of the individual processes that support operational concepts selected to achieve the enterprise intended strategy demands. Typical process management and ownership–related responsibilities include the following:

1. Specifying objectives (goals) and measures that relate to the objectives and targets to be achieved – these targets should be broken down into daily or weekly measures to enable continuous monitoring and management.
2. Communicating the objectives, measures and targets to the people executing the processes and, if necessary, providing rewards and incentives.
3. Monitoring and managing progress of the targets, and verifying whether the objectives and measures are still accurate and relevant.
4. Motivating staff to exceed objectives and deal with process disturbances.
5. Encouraging staff to identify bottlenecks and possible process improvements.

The management of process improvement – As management of process improvement, this aspect is concerned with the identification, development and roll-out of the benefits of "Process Improvement and Management."

Process improvement and management within an enterprise is about providing: focus (clearer perception) and integration (greater shared and streamlined) work knowledge and insight within the enterprise businesses. Although process improvement and management might come naturally to our minds, it is not an involuntary practice within enterprise businesses.

Regardless of the drivers and triggers, shown in Table 10.1 which is adapted from (Jeston & Nelis, 2008b), a "Process Improvement and Management" initiative, in a similar vein as a "Performance Measurement" initiative described in a

Table 10.1 Triggers and drivers for implementation of a process improvement and management initiative

Perspective	Trigger and drivers
Enterprise business	High growth – difficulty coping with high growth or proactively planning for high growth
	Mergers and acquisitions – they cause the enterprise business to 'acquire' additional complexity or require rationalization of processes. The need to retire acquired legacy systems could also contribute
	Reorganization – changing roles and responsibilities
	Change in intended strategy – deciding to change direction to operational excellence, product leadership or customer intimacy
	Intended strategic objectives or goals are not being met – introduction of process management, linked to organizational strategy, performance measurement and management of people
	Compliance or regulation
	The need for business agility to enable the enterprise business to respond to opportunities as they arise
	The need to provide the enterprise business with more control of its own destiny
Management	Lack of reliable or conflicting management information – process improvement and management and performance measurement and management will assist
	The need to provide managers with more control over their processes
	The need for the introduction of a sustainable performance environment
	The need to gain the maximum return on investment from the existing legacy systems
	Budget cuts
	The need for the ability to obtain more capacity from existing staff for expansion
Employees	High turnover of employees, perhaps due to the mundane nature of the work or the degree of pressure and expectations upon people without adequate support
	Training issues with new employees
	Low employee satisfaction
	The expectation of a substantial increase in the number of employees
	The wish to increase employee empowerment
	Employees are having difficulty in keeping up with continuous change and the growing complexity
Customers Suppliers Partners	Low satisfaction with service, which could be due to:
	High churn rates of staff
	Staff unable to answer questions adequately within the required timeframes
	An unexpected increase in the number of customers, suppliers or partners
	Long lead times to meet requests
	An organizational desire to focus upon customer intimacy
	Customer segmentation or tiered service requirements
	The introduction and strict enforcement of service levels
	Major customers, suppliers and/or partners requiring a unique (different) process
	The need for a true end-to-end perspective to provide visibility or integration

(continued)

Table 10.1 (continued)

Perspective	Trigger and drivers
Product and services	An unacceptably long lead time to market (lack of business agility)
	Poor stakeholder service levels
	Each product or service has its own processes, with most of the processes being common or similar
	New products or services comprise existing product/service elements
	Products or services are complex
Processes	The need for provision of visibility of processes from an end-to-end perspective
	Too many hand-offs or gaps in a process, or no clear process at all
	Unclear roles and responsibilities from a process perspective
	Quality is poor and the volume of rework is substantial
	Processes change too often or not at all
	Lack of process standardization
	Lack of clear process goals or objectives
	Lack of communications and understanding of the end-to-end process by the parties performing parts of the process
Technology	The introduction of new systems, for example CRM, ERP, billing systems
	The purchase of process management automation tools (workflow, document management, business intelligence), and the enterprise business does not know how to best utilize them in a synergistic manner
	Phasing out of old application systems
	Existing application system overlaps and is not well understood
	Introduction of a new IT architecture
	A view that IT is not delivering to business expectations
	The introduction of web services

previous chapter, is implemented in a social, economic and environmental context within the enterprise, and has intended or unintended positive and/or negative impacts. In much the same as with performance measures within an enterprise business, implementation of "Process Improvement and Management" initiative can be experienced in a positive or a negative manner depending on the maturity stage of the enterprise business.

Indeed, each and every function will be experienced quite differently in within an enterprise at the first or at the second stage of maturity, where leaders dictate or a command-and-control environment prevails. If employees perceive (through integration and context) that a process improvement and management initiative is in place to help them to become more effective and efficient, then the initiative will become a powerfully positive force in the enterprise.

Therefore, enterprise business executives, managers and leaders must address four aspects of paramount importance to making progress on moving the "Process Improvement & Management" initiative from its current maturity stage to "Continuous Improvement" maturity stage: Context of the initiative, Focus of specific technical content, Integration management of specific technical content, and Interactivity of the initiative, as shown in Fig. 10.1.

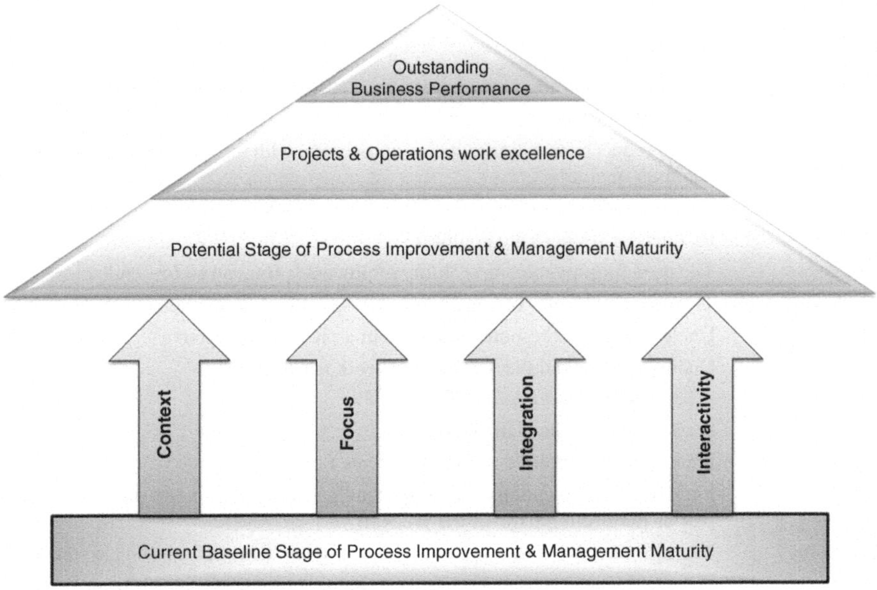

Fig. 10.1 Four keys to transforming "Process Improvement and Management"

At the base of the diagram shown in Fig. 10.1 is the current stage of "Process Improvement & Management" in your enterprise business – even though there might be some enlightened process improvement projects underway. The enterprise business might have a process improvement methodology, a process management methodology (perhaps even a technology-enabled business process management one), and dedicated specialists well-trained in the application of these methodologies.

At this basic stage, enterprises can take advantage of at least some of the functionality that process improvement and management has to offer. However, in order to tap into its real power, it is important to progress far beyond this baseline stage. As the diagram in Fig. 10.1 shows, the extent to which an enterprise business can make effective use of the four keys is the extent to which it can tap into the true prospective of potential stage of process improvement and management, which will enable "Projects & Operations work excellence," and which will, in turn, enable "outstanding enterprise business performance" – the ultimate goal of "Continuous Improvement" maturity stage.

10.3.1 Context of "Process Improvement and Management"

The context of a "Process Improvement and Management" initiative refers to the circumstances that form the setting for "Process Improvement and Management" event, statement, idea, constraints, social climate, or human factors, and in terms of which it can be fully understood and assessed. It is represented by everything else that surrounds and affects the specific technical content of the initiative, and it ultimately determines the effectiveness of implementation of any initiative within an enterprise. It reflects how process improvement and management is perceived by employees and consequently how they respond emotionally to it.

How employees respond to a process improvement initiative is largely a function of how the process is used. Moreover, the context of a "Process Improvement and Management" initiative can make the difference between employees being energized by initiative or employees just minimally complying with it, and even using it for their own personal benefit.

The factors that strongly affect the context of a "Process Improvement and Management" initiative are identical to the factors that strongly affect the context of performance measurement and management. As indicated in a previous section, these are: the climate within the enterprise, the "Process Improvement and Management" expectations, and the human factor.

The climate within the enterprise – As mentioned already, it is the prevailing "atmosphere" within the enterprise, the social-psychological environment that profoundly influences all behavior, and it is typically measured by employees' perceptions. The prevailing "atmosphere" is what best "defines" an enterprise to employees. It reflects perceptions on a variety of dimensions, including, among others:

1. The extent of formality (hierarchical structure) versus informality
2. Trust versus distrust (and cynicism) of employees
3. Open versus closed communication
4. Controlling versus collaborative decision making
5. Inward-looking versus outward-looking
6. Past focus versus future focus
7. Task-focus versus people-focus
8. Change versus rigidity
9. Risk-taking versus risk aversion

Enterprises at the "Continuous Improvement" stage of maturity are characterized by a prevailing "atmosphere" that is most conducive to improvement initiatives. These enterprises tend to be rated highly in such dimensions as openness, trust, honesty, collaboration, customer-focus, and flexibility.

The "Process Improvement and Management" expectations – It describe the practices, and the "rules" of conduct relative to process improvement and management within the enterprise. Although not always explicitly documented, and often unwritten, these expectations tend to reflect the enterprise's assumptions, its deeply-held beliefs about process improvement and management. For example, expectations will prescribe what types of processes are most important for improving. In most enterprise businesses today, critical processes are still much more highly valued than non-critical ones.

The Human Factor – This is the most critical component of the context of process improvement and management. Processes are executed either by people or by people supported by technology. The ideas and inspirations that guide and improve processes within the enterprise businesses come from its people. Indicators of process performance measures; i.e. the actual specific measures of process performance needed to decide on the required level of potential improvement, are of no value without human involvement. Drucker, writing in (1964), suggested that:

> *Business is a human organization, made or broken by the quality of its people. Labor might one day be done by machines to the point where it is fully automated. But knowledge is a specifically human resource. It is not found in books. Books contain information; whereas knowledge is the ability to apply information to specific work and performance. And that only comes with a human being, his brain or the skill of his hands.*

It is people who will ultimately determine the effectiveness of a "Process Improvement and Management" initiative. People bring knowledge, skills, attitudes, commitment, capabilities, and experience into their execution of processes.

Transforming and optimizing (production and/or service) processes is and remains a leading concern of enterprise businesses worldwide. Enterprise businesses and companies are launching process improvement and management initiatives to optimize their processes effectiveness and efficiencies. The first thing that enterprise business executives, managers and leaders must realize is how important it is to establish an environment conductive to transforming "Process Improvement and Management" behaviors. The context of processes must also be redesigned to make their execution more appealing. Traditional production systems, which treat employees as cogs in a machine, have been notoriously making employees less eager to work and commit to any improvement initiative.

10.3.1.1 Improving "Process Improvement and Management" Context

Improving the context of process improvement and management is one of the best investments an enterprise business can make, since the context affects all other aspects of any improvement initiative across the entire enterprise business. If the context is not transformed, then most people, if they use the improvement initiative at all, will just be "going with the flow" and will very likely also continue using the initiative for their own self-serving purposes.

To create a positive context for process improvement and management within enterprises at lower stage of maturity, enterprise business executives, managers and leaders must break with tradition, keeping in mind that the purpose of a "process improvement and management" initiative within the enterprise business is to provide clearer perception, greater shared and streamlined work knowledge and insight. Indeed, most enterprise business executives, managers and leaders consider their enterprise business to be a machine with employees as cogs. They create rigid structures with rigid rules and then try to maintain control by "pulling levers" and "steering the ship." Creating a positive context means breaking from the employee as cogs tradition. Encourage employees to be active, think and take initiatives, and enjoy their work. Here are some ways to make this happen:

1. Recognize the difficulty
2. Assess people attitudes
3. Demonstrate visible commitment
4. Keep employees productively busy

5. Allocate the resources
6. Create a climate of involvement and appreciation
7. Maximize employee input
8. Emphasize the importance of learning
9. Encourage productive social interaction

Recognize the difficulty – Transforming the context of "Process Improvement and Management" from its current baseline stage to its potential maturity stage requires a very significant pattern shift from the way things are currently done in most enterprise businesses. For others than those who specialize in it, process improvement and management is not something that most people want to do or feel that they do well. Process improvement and management is a habit that must be developed

Assess people attitudes – As enterprise executive, manager or leader, you should consider assessing existing attitudes in your enterprise business toward process improvement and management in order to gauge how difficult the journey will be. This will also help you to determine areas within the enterprise business that might be more receptive during the early stages of the journey, and to identify individuals who might be early adopters. It is not important that the entire population of the enterprise business be "fit for transformation."

A typical distribution of people attitudes is illustrated in Fig. 10.2. What this figure suggests is that only a small percentage of people in the organization (the right tail of the curve) will welcome a transformation effort and actively participate. Another small group (the left tail) will fight it actively. And the great majority – although they may nod and indicate their support – will be on the fence and waiting to see what is going to happen. Do not try to change those who, through blind ignorance, are clearly resistant to transforming the context of "Process Improvement and Management" – nothing is more frustrating than that. Look for those visionary managers, leaders and employees who "get it" and who are receptive – people who are likely to be the "early adopters" of transforming the context.

Demonstrate visible commitment to process improvement and management – Process improvement and management must be truly and authentically valued by those who lead it, or the remaining of the enterprise business populations will detect the lack of integrity. Therefore, it is important for the enterprise business executives, managers, and leaders driving the transformation of process improvement and management to become educated in the principles and practices involved in process improvement and management.

Keep employees productively busy – In a positive context, employees should leave work feeling that they accomplished something worthwhile. Do not allow them to be passive. Instead of letting them wait for assignments, for example, encourage them to use downtime to carry out self-improvement activities or ways to improve processes, hence activities, they are working on.

Allocate critical resources – The critical resources are assets such as the people, technology, products, facilities, equipment, channels, and brand required to deliver the value proposition to the targeted customer. The focus here is on the critical elements that create value for the customer and the enterprise business, and the way

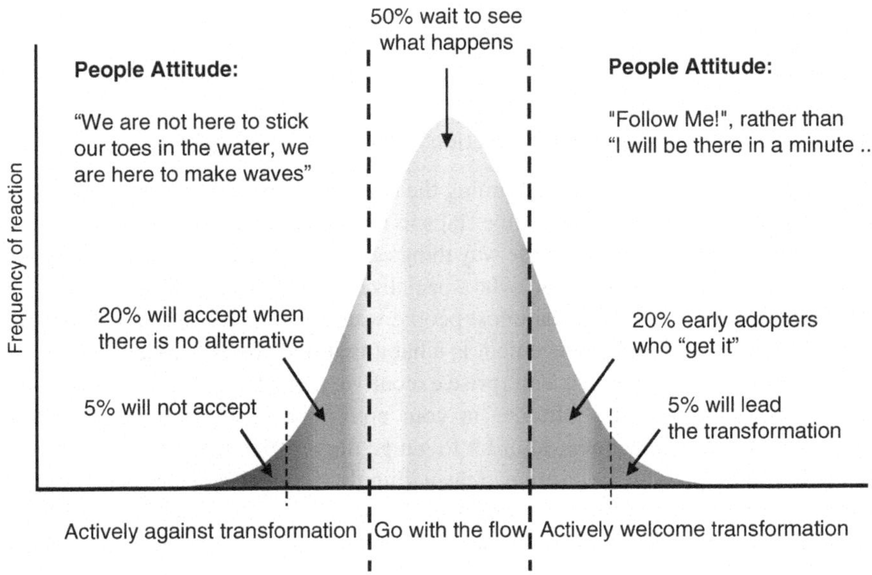

Fig. 10.2 Typical reaction to "Process Improvement and Management" transformation

those elements interact. Every enterprise business also has generic resources that do not create competitive differentiation.

Although process improvement and management has a cost associated with it, if done right, it delivers enormous value to projects and operations work critical to achieving the enterprise intended strategic demands. Do not starve the transformation initiative before it has the opportunity to take root. Allocate the critical resources, including education and training, necessary for making the transformation of process improvement and management a reality.

Create a climate of involvement and appreciation – Most traditional production systems provide a low level of positive recognition. Well thought out expressions of appreciation are powerful drivers of creating and enhancing positive contexts. As the context of process improvement and management progresses and the maturity stage increases, more and more people in the organization will become involved in initiative (from the lowest organizational level to the highest organizational level) and will begin to experience its positive side.

Involvement starts with the "early adopters," but it increases as additional process improvement opportunities are identified, and employees experience personal involvement in using the improved processes to achieve the objectives of projects and operations work they are assigned to. As the transformation process continues, employees will develop more ownership in process improvement and management:

1. *Maximize employee input* – Employees are a great source of ideas. And they will be committed to an enterprise business willing to listen to them.

2. *Emphasize the importance of learning about and from process improvement and management* – Learning from process improvement and management should be considered one of the key outcomes of the transformation initiative.
3. *Encourage productive social interaction during process improvement* – Interaction enhances communication and cooperation.

Transforming the context of "Process Improvement and Management" from its current baseline stage to its potential stage may take time, and we encourage enterprise business executives, managers and leaders to always start on a small prototype scale. Within the enterprise business, some functional areas are likely to be more receptive to this transformation than others.

We also encourage enterprise business executives, managers and leaders to start with the functional area of the enterprise business which is more receptive to this transformation, as a "prototype of concept." Once the concept is shown to be effective, the remaining functional areas within the enterprise business will be more receptive to broader transformation of "Process Improvement and Management."

10.3.2 Focus of Specific Technical Content

The second aspect of paramount importance to making progress on the development and implementation of a "Process Improvement and Management" initiative is "Focus of specific technical content." Within an enterprise business, there is a variety of processes that can be identified and a lack of focus can only leads to a waste of resources that should be assigned on the critical few high leverage projects and operations work drivers of the most important results of the intended strategy. When every process is important, there is nothing that is most important. Focusing on the right process creates enormous leverage for the enterprise business.

The purpose of focusing the specific technical content of a "Process Improvement and Management" initiative is to differentiate between the critical few high leverage processes associated with the operations and projects work that are drivers of the most important outcome of the enterprise intended strategy and the variety of other processes, the trivial many, that permeate every area of an operations and projects work within the enterprise and keep the enterprise running.

Focused specific technical content is about being effective, getting the right process captured, improved and standardized. In contrast, efficiency – which is too often the primary focus of enterprise managers – is about minimizing resources that support the work being completed, as enterprise businesses desperately try to drive out cost. While efficiency is important, effectiveness of specific technical content must come first. Indeed, there is no value in executing efficiently those processes that should never be executed at all. Anything that is not effective is waste, and reducing waste is, in turn, a key to increasing efficiency.

Successful enterprise businesses have operational and managerial processes that allow them to deliver value in a way they can successfully repeat and increase in scale. These may include such recurrent tasks as training, development, manufacturing, budgeting, planning, sales, and service. Such critical processes also include an enterprise business' rules, metrics, and norms. These four elements form the building blocks of any business. The customer value proposition and the profit formula define value for the customer and the enterprise business, respectively; critical resources and critical processes describe how that value will be delivered to both the customer and the enterprise business.

The simple approach to developing a focused specific technical content for process improvement and management is:

1. To capture the critical processes resulting from the enterprise business QFD alignment, as explained in the "Alignment" chapter, and
2. To develop the process architecture for the enterprise processes.

The critical processes and their documentation will form the basis for the specific technical content.

Capturing the critical processes resulting from the enterprise QFD alignment defines the enterprise process models. It also provides a structured approach to the enterprise business process models and ensures that processes are effectively and efficiently contributing to achieve operations and projects work critical to accomplish the enterprise intended strategy. Not every process within the enterprise business contributes towards the achievement of operations and projects work critical to accomplish the enterprise business intended strategy. Critical processes are the ones that do.

The process architecture on the other hand will ensure that all the relevant information, which consist of the foundation and guidelines for the process review and improvement, are made explicit and can be referred to. The process architecture is much more than a process model. It comprises a process model plus the objectives, principles, policies and guidelines that are the foundation for reviewing or creating new processes. Good process architecture for the enterprise will guarantee minimum time and effort for its use and provide a means of communicating, specifying and agreeing clear objectives for reviewing or creating new processes. All well-defined and well-managed processes have some common characteristics:

1. They have someone who is held accountable for how well the process performs (the process owner).
2. They have well-defined boundaries (the process scope).
3. They have well-defined internal/external interfaces and responsibilities.
4. They have documented procedures, work tasks, and training requirements.
5. They have performance measure and feedback controls close to the point at which the activity is being performed.
6. They have customer-related performance measures and targets.
7. They have known cycle times.
8. They have formalized change procedures.
9. They know how good they can be.

Within an enterprise business, processes are generally clustered into several major categories and hierarchies related to operational performance concepts (described in the previous section) necessary to achieve the enterprise business intended strategy demands.

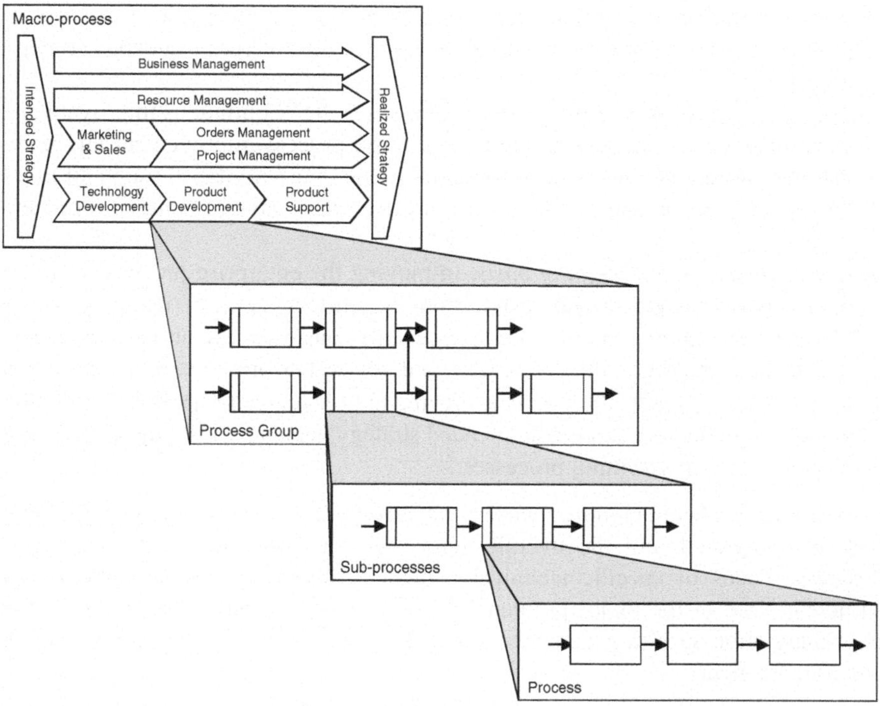

Fig. 10.3 Process clustering

These clusters – process model, process groups, and sub-processes – and their constituent processes, as illustrated in Fig. 10.3, serve as guide to apply appropriate knowledge, skills and resources during the course of projects and operations work. Furthermore, the application these process groups to projects and operations work is often iterative and their constituent processes are repeated and revised during the work.

Examples of process groups are: Corporate Governance Process, Strategic Planning and Management Process, Purchasing Process, Technology Development Process, Product Development Process, Manufacturing Process, Advertising and Marketing Process, Sales Process, Accounting and Finance Process, Technical Support Process, etc.

10.3.2.1 Improving the Focus of Specific Technical Content

Improving the focus of specific technical content aims at increasing the use of critical and high-leverage processes necessary to complete projects and operations work. Here are some ways to make this happen:

1. *Focus on the enterprise intended strategy demands.* Enterprise businesses at the "Continuous Improvement" stage of maturity constantly reassess and recalibrate their processes against their business model and intended strategy. Then, they

focus on the processes that are most crucial for achieving the objectives of projects and operations work critical to realizing the enterprise intended strategy demands.

2. *Encourage innovation.* Enterprises businesses at the "Continuous Improvement" stage of maturity encourage innovation in the process improvement and management system. Knowing the importance of viewing activities through different lenses, they encourage employees to discover new ways of achieving process tasks.
3. *Review/revise on an ongoing basis.* In moving the enterprise process improvement and management dimension from its current stage of maturity to the "Continuous Improvement" stage, enterprise business executives, managers and leaders must continually review and revise performance of processes in terms of how valuable they are in achieving high-leverage projects and operation work to realizing the enterprise intended strategy demands. They must change or discard poorly performing processes.

As processes become more focused and clustered, it is even more essential that they be integrated into the overall framework and structure of the enterprise business. Focus of specific technical content is necessary, but it alone is not sufficient. One of the major problems in most enterprise businesses today is the poor integration specific technical content associated with "Process Improvement and Management."

10.3.3 Integration Management of Specific Technical Content

In some enterprise businesses, functions and their processes operate so independently that there is virtually no connection between them at all. Poorly integrated processes allow managers to pursue their own or departmental interests ahead of those of the enterprise business or its shareholders. The third aspect of paramount importance to making progress on the development and implementation of a "Process Improvement and Management" initiative is "Integration management of specific content" across the enterprise business; i.e., the relationships and overall trade-offs and balance among several different factors to create an optimal configuration of process clustering across the enterprise business.

As powerful as individual critical processes can be to create value, they can also become ineffective if they are not integrated into a framework that shows how they are related to other processes. It is the relationship and overall trade-offs and balance that will yield consistent, ongoing value creation over the long term. The bottom line message to enterprise business executives, managers and leaders is: if "Process Improvement and Management" is going to have a truly transformational long-term impact, it must reflect the interconnectedness and holism of the total system – and integration across the enterprise business.

Integration management of specific technical content includes characteristics of unification, consolidation, articulation and integrative actions that are crucial to process execution, successfully achieving operations or projects work objectives. It deals with the relationship and overall trade-offs and balance; i.e. making choices about where to concentrate resources and effort execution of a specific process, anticipating potential issues or process shortcomings, dealing with these issues before they become serious.

Clusters of processes and their constituent processes are often presented as discrete components with well-defined interfaces; while in practice they overlap and interact in complex ways, the description of which extends beyond the scope of this chapter. Integration management is primarily concerned with effectively integrating the constituent processes among the process clusters that are required to accomplish projects and operations work objectives within the defined procedures of the enterprise business.

The need for integration management of specific technical content becomes evident in situation where individual discrete components interact. For example, a cost estimate of a project work needed for a contingency plan involves integration of the planning processes associated with cost management process, time management process, and risk management process. When additional risk associated with the use of alternative resources to support the work been completed by the process are identified, then one or more of the planning processes must be revised.

The integrative nature of processes can be better understood if we think of the other processes been executed while completing operations or projects work. Most enterprise businesses are composed of functional silos. They typically evolve from small simple ones to large complex ones. As separate functions emerge, each function wants to own its share of processes and resources. Functional boundaries become strong and more entrenched. Traditional disciplinary thinking creates and reinforces individual departmental factions. People tend to "see" the enterprise business through their own functions, roles, and processes, i.e., by their job descriptions and what they need to do to achieve success.

Therefore, we encourage enterprise business executives, managers and leaders to continually remind people in different functions within the enterprise to look beyond their processes. Cross-functional communication will help. But whatever steps are taken, none of them will be truly effective – at least not for long – unless the integration management of specific technical content of "Process Improvement and Management" is performed. Without integration management of specific technical content, the enterprise business will inevitably be at risk of operating at cross-purposes, often imperceptibly, wasting resources that could be focused on mutually creating real value.

There are actually two types of integration management of specific technical content: vertical and horizontal. Vertical integration involves the connection between intended strategy and critical processes up and down through the enterprise business. Horizontal integration is the connection of processes across the enterprise business functions and processes. Both of these forms of integration management can be obtained by mapping and capturing the critical processes associated with the enterprise QFD alignment described in a section above. Using the QFD alignment as framework and the process architecture developed for the enterprise processes to focus the specific technical content for process improvement and management, the enterprise business can further purchase a process-modeling technology tool to facilitate the integration management and enable better enterprise-wide processes execution.

We must clarify at this stage that the integration management of specific technical content is about establishing a holistic process improvement and management system that must exist within a positive context; it is unequivocally not primarily about technology. Of course any breakthrough in enterprise process improvement and management will eventually require technology, but technology "solutions" are not real solutions unless the social enablers are in place. Technology should support the holistic process improvement and management system – it is not the system!

As we have illustrated throughout this "Process Improvement and Management" section, significant process improvement and management can be achieved without the use of a technology process-modeling tool or software. The technology is helpful for enabling people to better deal with the complexity of enterprise processes and the proliferation of data, and to assist in mining data for insights. It can assist in performing some tasks that people cannot perform effectively themselves, or perform inefficiently. For instance, the technology can automate data collection process; reduce data handling errors; perform intricate analytics (including modeling "what-if" scenarios); enable simulation and predictive modeling; present data in virtually any form, with impressive visualization capabilities, even customized for each stakeholder group; zoom in on process detail, zoom out to see the big picture.

Technology can definitely reduce human intervention where it does not add unique value, and prepare information so that it is ready for the kind of interpretive activity that people can do best. One of the keys to success for enterprise businesses is to recognize when to rely on technology, and when to recognize and work around its limitations. It is not a matter of choosing between technology and people; it is a matter to using each appropriately.

10.3.3.1 Improving Integration Management

Improving the integration management of specific technical content is not primarily about isolated processes, but about increasing progress toward one integrated process improvement and management system, including replacing the functional "silo" processes and their data repositories. However, nothing is more difficult than overcoming the forces of functional parochialism and data politics.

As activities become more complex, enterprise businesses need to divide up functions. Operating a complex enterprise business requires excellence from many functional areas: human resources, production, sales, marketing, accounting, finance, inventory management, compliance, risk management, and more. Enterprise businesses also need people to lead and manage those functions.

As responsibilities grow for each department, so does the pressure and accountability to meet local goals. At some point, out of necessity and in response to daily pressures and demands, the managers and leaders of those departments may start to focus much more on their functional goals than on the enterprise overall goals.

When this happens, making sure that their part of the process is done correctly may become all that really matters, even at the expense of mission success. The department or function may not only lose sight of the connection between its work and organizational outcomes; it may even stop caring about what happens outside its silo. The department or function then defines its world by the piece, not the puzzle. This condition is referred to as functional parochialism.

Parochialism develops when a group views the world strictly through the lens of its functional goals, and it judges the relative importance of other activities by the way they affect the group's objectives. Parochialism limits the group to a narrow reference point – ultimately, everything is viewed from that filtered local perspective. Furthermore, if an enterprise business is already suffering from functional parochialism, then competition over resources can become intense. When this occurs, managers and leaders may start to lose sight of what is best for the business and focus just on what is best for their small part of the organization. Loss aversion kicks in, creating a natural tendency to protect and maintain excessive control over headcount and resources. If allowed to continue, the result may be territorialism, or the exertion of control over one's silo to an extent that harms others in the same business.

When a functional parochialism situation occurs within an enterprise business, as it often does, moving forward with the integration management of specific content is enabled by the following factors:

1. Holistic view
2. Development and use of process frameworks
3. Cross-functional processes
4. Cause-and-effect understanding
5. Ongoing strategy alignment

Holistic view – Enterprise businesses at the "Continuous Improvement" stage of maturity are working hard to take a holistic, "big picture" view of process improvement and management, increasingly improving process constructs that reflect a broader understanding of all essential value creation activities and being more aware of their trade-offs.

Development and use of process frameworks – Enterprise businesses at the "Continuous Improvement" stage of maturity realize that isolated processes must be integrated into larger "process frameworks." These frameworks include both vertical integration (the connection of processes up and down through the enterprise business) and horizontal integration (the connection of processes across the enterprise business functions). Enterprise businesses at the "Continuous Improvement" stage of maturity are finding that they are often dealing with abstract concepts, exploring in areas where many do not feel very comfortable, and adopting new processes in areas of considerable uncertainty. Process frameworks enable enterprise businesses to address these issues.

Cross-functional processes – Most of what adds value in enterprise businesses today is cross-functional. In enterprise businesses at the "Continuous Improvement" stage of maturity, cross functional processes contribute greatly toward breaking down long established functional silos. New cross-functional processes, which are viewed as a key to collaboration across the enterprise business, are being regularly adopted.

Cause-and-effect understanding – It is through understanding the relationships and the trade-offs among process factors that enterprise businesses can gain valuable predictive insights from their process improvement and management systems. They realize that the key to the transformation of process improvement and management is to "understand first; execute second." One of the indicators of process improvement and management maturity is that the causal relationships are being frequently being hypothesized and tested. Too many organizations hypothesize, but do not test. Insight into important cause-and-effect relationships and healthy skepticism about these relationships are features of process improvement and management maturity.

Ongoing intended strategy alignment – In enterprise businesses at the "Continuous Improvement" stage of maturity, there is ongoing commitment to increasing deeper understanding of the key strategic drivers of the enterprise business success. This is not a one-shot alignment exercise, but an ongoing process. One of the major contributions of QFD alignment has been to gain greater awareness of the importance of increasing the alignment on the enterprise intended strategy. However, as you have seen, this is just one element of process improvement and management maturity.

10.3.4 Interactivity of "Process Improvement and Management"

The fourth aspect of crucial importance to making progress on the development and implementation of a "Process Improvement and Management" initiative is "Interactivity." It represents the social communicative aspects of "Process Improvement and Management" which occurs through the search for shared knowledge or understanding of processes.

Because processes need to be integrated across an enterprise business, functions and people supporting the work been completed by those specific processes within the enterprise business need be become more interactive through dialogue. This interaction around them is what will turn the development and implementation of a "Process Improvement and Management" initiative within the enterprise into a transformational and effective reality.

Unfortunately, as we mentioned with in the performance measurement chapter, very few people are skilled at dialogue, and very few enterprise businesses currently have a strong capacity for dialogue. In fact, dialogs in most enterprise businesses are suppress in favor of debates, the more formal and adversarial processes which are antithetical to dialogues, because the purpose is for one individual to win an argument.

In order to take advantage of the interactivity that should occur at every stage of the development and implementation of a "Process Improvement and Management" initiative, enterprises business executives, managers and leaders must endeavor to enhance and consolidate a positive context of "Process Improvement and Management" initiative through dialogue.

Dialogue thrives on openness, honesty, and inviting multiple viewpoints, as we indicated already. In dialogue, diversity of perspective is almost always good – whether it be functional, cross-functional, local, global, systemic, etc. The more perspectives involved, the richer the dialogue can be around the specific technical content of a "Process Improvement and Management" initiative. Dialogue as interactivity should incorporate: learning, understanding, defining, listening, modeling, hypothesizing, balancing, linking, and integrating.

Although most enterprises have a long way to go on the development and implementation of a "Process Improvement and Management" initiative, some enterprises at the "Continuous Improvement" stage of maturity have been identified as being more effective than most in "Process Improvement and Management." These enterprises at the "Continuous Improvement" stage of maturity are more successful because of how much more effectively they use processes as a critical part of managing and doing projects and operations work on a continuing basis.

10.3.4.1 Improving "Process Improvement and Management" Interactivity

Enterprise businesses that are improving the interactivity of processes are well on their way to the "Continuous Improvement" maturity stage. Most of them have already improved through some degree of transformation of the context of the initiative, focus of specific technical content, and integration management of specific technical content. Now, they are looking for new and better ways to "socialize" process improvement and management. These enterprise businesses are using interaction to develop and continually review new process improvements, supplemented by the appropriate use of technology.

Enterprise businesses at the "Continuous Improvement" stage of maturity have typically established a "social architecture" to promote discussion of aspects of process data and information. This formal or informal structure enables enterprise operation managers and project leaders to carry on regular dialogues about process improvement and management issues.

Process improvement and management is built into the social fabric of the enterprise business, and is no longer just a program or an add-on. Most of the mistakes and shortfalls in attempts at the transformation of process improvement and management from the baseline maturity stage to the "Continuous Improvement" stage of maturity have, in fact, been due to a lack of interactivity. The factors which contribute to transform the interactivity of "Process Improvement and Management" include the following:

1. Frequent Interactivity
2. Effective and robust dialogue
3. Incremental and ongoing review and improvement of process frameworks
4. Collaborative learning
5. Appropriate use of technology

Frequent Interactivity – Operation managers and project leaders realize that process improvement and management is primarily about clearer perception, deeper understanding, and greater shared insight, knowledge, and wisdom.

Effective and robust dialogue – As Larry Bossidy and Ram Charan have pointed out, how people talk to each other absolutely determines how well the enterprise business will function. A lot of this interaction occurs through dialogues. Regular dialogue meetings within teams, groups, and between functions will help to integrate functions and lead to higher levels of collaboration and performance in process improvement and management.

Incremental and ongoing review and improvement of process frameworks – Enterprise businesses at the "Continuous Improvement" stage of maturity realize that what is optimal at a particular time might not be optimal in a month or a year, so the refinement of the process frameworks must be continuous. Processes must be continually calibrated and realigned with strategy, and then integrated across the entire enterprise businesses.

Collaborative learning – Most learning from process improvement and management is "collaborative action learning." Through using and continuously improving processes, people in enterprise businesses at the "Continuous Improvement" stage of maturity are engaging in both single-loop and double-loop learning, and they are not afraid to challenge the traditional assumptions about process performance and existing process frameworks. There are regular "dialogue" meetings between functions to discuss existing processes, develop actions plans, review process frameworks, and consider process improvement issues. Enterprise businesses are finding that dialogues about their process frameworks will help identify cross-functional processes that will make a transformational difference to the enterprise. Nothing will break down the traditional functional barriers like collaborative learning through cross-functional processes.

Appropriate use of technology – In enterprise businesses at the "Continuous Improvement" stage of maturity, technology is viewed as an enabler of a robust process improvement and management system, but not the system itself. The emphasis is on automating routine processes and administrative functions, and on performing advanced analysis and reporting, but not replacing the uniquely human capabilities or detracting from the social aspects of the transformation of process improvement and management. Care is taken that technology facilitates interactivity of process improvement and management, and does not diminish it.

10.4 Conclusion

We have indicated that the purpose of this section on "Process Improvement and Management" is not about doing the systematic methodology for process improvement; it is about creating an optimal environment for its effective use of process improvement and management.

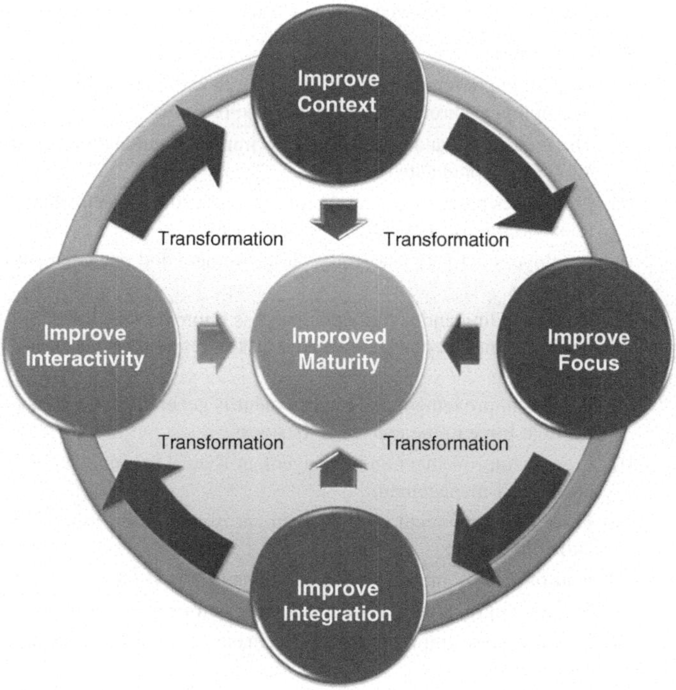

Fig. 10.4 Transformation process for "Process Improvement and Management"

Attaining the optimal environment requires a specific and intensive set of actions – a transformation process depicted in Fig. 10.4, and adapted from Dean Spitzer, as a continuous improvement loop; progressing from improving context, to improving focus, to improving integration, to improving interactivity – the four aspects of paramount importance to making progress on moving the "Process Improvement & Management" initiative from its current maturity stage to "Continuous Improvement" maturity stage.

This transformation process for "Process Improvement & Management" does not necessarily occur in this particular order, it is important that all four aspects of crucial importance be improved incrementally and on an ongoing basis. While the transformation cycle is occurring, maturity (in the center of the diagram) is also increasing. The transformation must be ongoing, or it will stop when the initiative is deemed implemented. The achieved maturity can be assessed in an ongoing basis the maturity assessment questionnaire given in Table 10.2 below (adapted from (Spitzer, 2007)).

The questionnaire is divided into four parts, one for each of the key aspects discussed in this chapter: Context, Focus, Integration, and Interactivity. The assessment score gives an indication of the extent to which the enterprise business has progressed overall and in each key aspect of the transformation. Although comparing the enterprise business' total score to the maximum score of 250 will give

Table 10.2 "Process Improvement and Management" maturity questionnaire

	#	Observation	Rating
Context maturity	01	Process improvement and management is widely used by all levels of employees throughout the enterprise business	
	02	The importance and value of process improvement and management are widely appreciated	
	03	Employees perceive process improvement and management as relevant, timely, and actionable in their jobs	
	04	Employees actively use process improvement and management in their jobs	
	05	Understanding and acting upon process improvement and management outcomes are viewed as key responsibilities of all employees	
	06	Process improvement and management is generally viewed as a positive force in the enterprise business	
	07	Process improvement and management is used to empower and enable self-management	
	08	Process improvement and management is rarely used to identify culpability	
	09	Fear of process improvement and management is low	
	10	Process improvement and management is trusted	
	11	Use of process improvement and management for self-interests is low or nonexistent	
	12	Process improvement and management outcomes are discussed openly and honestly	
	13	Employees are educated about process improvement and management	
	14	Employees are given the time and other resources they need to improve processes	
Focus maturity	15	This enterprise business addresses critical and high-leverage processes and not those that don't matter	
	16	Process improvement and management accurately reflect the most critical aspects of the operations and project work for achieving the enterprise intended business strategy	
	17	Processes are regularly reviewed and revised or eliminated (as appropriate)	
	18	This enterprise business has the right number of processes (not too many nor too few)	
	19	Routine processes are reduced when new high leverage processes are added	
	20	Routine processes are being increasingly automated	
	21	Progress is being made in capturing and improving processes associated to the measures of intangible assets	
	22	Experimentation with emergent processes is encouraged	
	23	Transformational and improved processes are being widely adopted and used	

(continued)

Table 10.2 (continued)

	#	Observation	Rating
Integration maturity	24	There is a holistic approach to process improvement and management across the enterprise business	
	25	Processes are becoming more integrated	
	26	Employees understand the cross-functional implications of their processes	
	27	Cross-functional processes are developed and used	
	28	There is increasing understanding of the relationships and trade-offs between processes	
	29	There is widespread commitment to understanding the causal relationships among processes	
	30	Integrative process frameworks are developed and used	
	31	Ongoing effort is being made to align process frameworks with intended strategy, and keep them aligned	
	32	Progress is being made toward creating one integrated enterprise-wide process improvement and management system	
	33	Process improvement and management integration efforts have enterprise-wide leadership	
Interactivity maturity	34	There is widespread and frequent interaction throughout the enterprise business about process improvement and management	
	35	Frequent interactivity occurs regarding the selection of processes	
	36	Developing and revising process frameworks are highly interactive	
	37	Insights from process improvement and management information are discussed in many forums	
	38	The organization places a high priority on learning from process improvement and management	
	39	Time is made available to learn from process improvement and management	
	40	There are frequent and high-quality dialogues about process improvement and management	
	41	Executives are deeply engaged in process improvement and management related dialogues	
	42	Process frameworks are continually and interactively reviewed and revised when appropriate	
	43	Interpretation of process outcomes is as highly valued in this enterprise business as data collection and analysis	
	44	Collaborative cross functional learning from process improvement and management occurs throughout the enterprise business	
	45	Revealing questions are constantly being asked about process improvement and management	
	46	Process improvement and management experiments and pilot projects are occurring throughout the enterprise business	
	47	The enterprise business has effective social mechanisms for translating process improvement and management outcomes into appropriate actions	

(continued)

Table 10.2 (continued)

#	Observation	Rating
48	The capability of the enterprise business for converting process improvement and management outcomes into actionable insight is high	
49	The enterprise business is effective at sharing insights from process improvement and management	
50	Technology is being used appropriately to support interactivity around process improvement and management	

Use the standard five-point rating scale:

5 = Strongly agree; 4 = Agree; 3 = Neither agree nor disagree; 2 = Disagree; 1 = Strongly disagree

Interpretation key:

Highest score is 250

210–250: Your enterprise business may just be at the cutting edge of process improvement and management, or you may be kidding yourself about how well your enterprise is doing. Very few enterprise businesses score this high

160–209: Your enterprise business is doing well, with some room for optimization. Focus on those areas among the four determining factors where your enterprise business is weakest for improvement

105–159: Your enterprise business is in dangerous territory, because it got here by scoring "3" in almost all areas, which is not very good. You might conclude that your enterprise business is average. In fact, it is probably shining in one or two contextual areas and doing poorly in the others

50–104: This is worrying; these scores are typical of enterprise businesses at low stage of maturity, authoritarian and where employees are viewed as expenses, not assets. Your enterprise business has major problems, and it is best to focus on one weak area at a time. The good news is that you have got a great opportunity to optimize process improvement and management within your enterprise business. We suggest, perhaps counter intuitively, that your enterprise business begins with addressing the context of process improvement and management initiative to build credibility

managers and leaders some idea of the "Process Improvement and Management" maturity, the primary purpose of this assessment is to help make the improvement initiative more visible, not to provide a static measure of the current stage of "Process Improvement and Management" maturity.

By administering this assessment over time, as "Process Improvement and Management" is being transformed, from its current maturity stage to the "Continuous Improvement" maturity stage, operations managers and projects leaders should be able to discern improvements. More importantly, this assessment questionnaire should be used for diagnosis and to foster dialogue about crucial "Process Improvement and Management" issues.

11 Sustainability

Sustainability relates to the degree to which an enterprise business incrementally and in an ongoing basis creates value to its customers and shareholders, captures value from its diverse assets (tangible and intangible), and attracts investors. Enterprise businesses that do not create value, by definition, destroy it; and unprofitable enterprise businesses are wasting both the money of their shareholders and the enterprise resources and assets.

In today's era of unprecedented change, complexity, volatility, and risk – where everything seems to be moving at extremely fast speed, there is very little room for error. The business imperative is not anymore just to perform excellently, but to perform excellently consistently. Sustaining success in today's hyper-competitive marketplace is an ultimate challenge for any enterprise business.

11.1 What Is Value Creation?

Nothing is more important today than the concept of "value creation and value capture," which it has become the benchmark criterion for success. Ultimately, the most successful enterprise businesses understand that the purpose of any business is to create value their customers, employees, and shareholders, and that the interests of these three groups are inextricably linked. Therefore, sustainable value cannot be created for one group unless it is created for all of them.

As illustrated in Fig. 11.1, the first focus of an enterprise business should be on creating value for its customers, but this cannot be achieved unless the right employees are selected, developed, and rewarded, and unless shareholders receive consistently attractive returns of their investments. An enterprise business creates value when the benefits provided to its stakeholders, customers, and employees exceed all the costs incurred.

For the customers, it entails making products and providing services that customers find consistently useful. In today's economy, such value creation is based typically on innovation and on understanding unique customer needs with ever-increasing speed and precision.

A. Van Aartsengel and S. Kurtoglu, *A Guide to Continuous Improvement Transformation*, Management for Professionals, DOI 10.1007/978-3-642-35904-0_11,

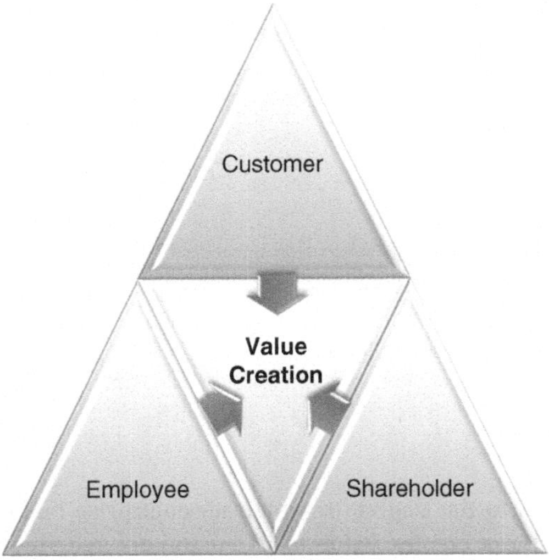

Fig. 11.1 Customer-employee-shareholder value triangle

But enterprise businesses can innovate and deliver outstanding services and products only if they tap the commitment, energy, and imagination of their employees. Value must therefore be created for those employees in order to gain their commitment and enable them. Value for employees includes being treated respectfully and being involved in decision-making. Employees also value meaningful work; excellent compensation opportunities; and continued training and development.

11.2 How Value Is Created and Destroyed

Within enterprise businesses, value is created when the right actions are taken and the right investments are made. Value is destroyed when the wrong actions are taken and the wrong investments are made. Value leaks or evaporates when nothing is done.

The right actions, within any enterprise business, are defined by the balance between two needs:

1. The need to look backward in order to maintain the existing business and its current customers, and
2. The need to look forward in order to explore and achieve performance breakthroughs and to identify and attract new customers and new sources of value.

Achieving this balance, as we indicated throughout the previous sections of this chapter, requires extraordinary capabilities and specific and intensive actions along eight determining factors of strategic management:

1. Leadership
2. Culture and Values
3. Strategic Planning and Management
4. Performance Measurement

5. Performance Management
6. Alignment and Commitment
7. Process Improvement and Management
8. Sustainability of Initiatives

When the right actions are taken within an enterprise business, value creation (and the avoidance of value destruction) occurs through the synergy that diverse critical resources and capabilities provide in achieving a balance of the eight determining factors. The larger the enterprise business is, the more important the synergy that diverse resources and capabilities provide within the enterprise business. Moreover, anybody in an enterprise business, at any organizational level, can contribute to creating value or destroying it. Employees who waste time or are otherwise unproductive, for example, destroy value.

The success of an enterprise business at the "Continuous Improvement" stage of maturity is based on doing the right things right and the enterprise business progressively becoming more expert. It learns successful ways of doing things. It finds out what its customers and shareholders like and gets good at delivering those things. It develops its technological expertise. It uses recipes which work and it becomes efficient. And it becomes effective. All this is by no means automatic, but happens as a result of deliberate management initiatives along the eight determining factors discussed throughout the previous chapters.

Failure in one of these eight determining factors results in unbalance and eventually value destruction. Value destruction, which results from taking the wrong actions, is often driven by poor decision-making, individual advocacy and self-interest, and not by what creates measurable long term value. Thus enterprise business executives and managers should, first and foremost, capture "the story" of value creation for their particular business.

Creating value for shareholders means delivering consistently high total returns on their investment capital. This generally requires both strong revenue growth and attractive profit margins. These, in turn, can be achieved only if the enterprise business delivers sustained value for its customers.

In business practices, value creation is typically measured by profitability in the short-term and long term growth. In order to achieve these goals, an enterprise business must establish, as a result of long-term oriented analyses, a process for continually developing and delivering a steady stream of products and services that offer unique and differentiated benefits to a chosen set of customers.

One example of a long-term oriented method of analysis which is based on the extrapolation of performance trends is the gap analysis. This form of analysis looks at performance already in existence and at products/services in development and predicts the rate of turnover or contribution margin they will achieve. Because of the product/service life cycle, turnover or contribution margin tend to reduce over time. By comparing projected future figures with what would be necessary to ensure the survival of the enterprise, a crucial performance gap can be identified.

Figure 11.2 shows how this is done. Just because an enterprise business has created value for its shareholders in the past does not mean that it will be able to continue to create value in the future. Long-term value creation is a challenge that

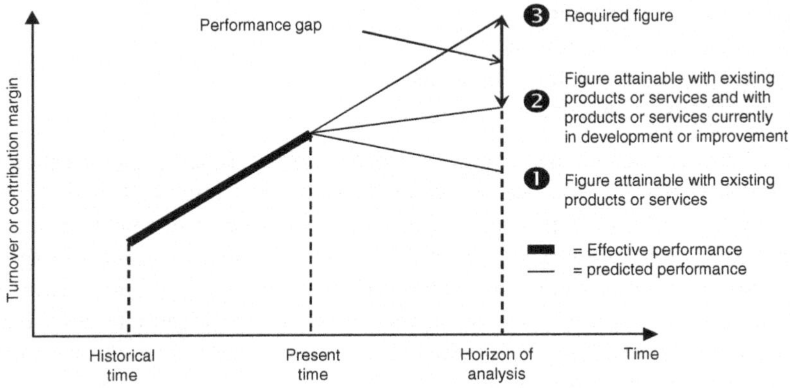

Fig. 11.2 Performance gap analysis

requires extraordinary and focused capabilities on the enterprise primary value-adding activities.

While gap analysis identifies the performance gap which needs to be bridged, it offers no clues as to how this can be done. Ansolff's analysis approach provides a practical framework for bridging this performance gap.

Ansolff's analysis simplifies the competitive position of an enterprise by defining two dimensions: The enterprise products/services and its markets. If we consider Ansoff's matrix shown in Table 11.1 we see that the gap in turnover or contribution margin can be closed with an improved market penetration, with new customers for existing products/services, with new products/services for existing customers, or with new products/services for new customers.

However, Ansoff's matrix cannot help an enterprise to decide which of these alternatives is to be preferred and to be attempted (Ansoff, 1965, 1976).

Improved Market Penetration – It occurs when an enterprise business chooses to enter an existing market with its current lines of products or services. It starts with the existing customers of the enterprise. This approach is used by enterprise businesses in order to increase sales without drifting from the original product-market intended strategy. Enterprise businesses often penetrate markets in one of three ways:

1. By gaining competitors customers;
2. By improving the product quality or level of service;
3. By attracting non-customers of the products or convincing current customers to use more of the enterprise product or service, with the use of marketing communications tools like advertising etc.

The market penetration approach to closing the performance gap is important for enterprise businesses because retaining existing customers, especially high lifetime value customers, is cheaper than attracting new ones. It is also used when the enterprise business' average cost of producing and distributing products or providing services decreases as the size of its operations increases. To proceed with a market

Table 11.1 Ansoff matrix

	Existing products/services	New products/services
Existing markets	Improved market penetration	Product/services diversification
New markets	Market diversification	Lateral diversification

penetration approach the enterprise business must hold a defendable competitive position to avoid likely counter-action from its competitors.

Market Diversification – It occurs when an enterprise business chooses to move beyond its immediate customer base towards attracting new customers for its existing products or services. Its numerous implementations often involve:

1. Development of new geographical markets;
2. Addition of new distribution channels;
3. Adoption of different pricing policies to attract different customers;
4. Creation of new market segments.

Products/service diversification – It occurs when an enterprise business chooses to introduce new products to existing customers. With this approach, the enterprise business chooses to develop modified products that appeal to existing customers in order to encourage them to spend more on these products. Note that product/service diversification refers to significant new product/service developments and not minor changes in an existing product/service of the enterprise business. The reasons that justify the use of this approach to closing performance gap include one or more of the following:

1. To use the enterprise business excess production capacity;
2. To counter competitive entry;
3. To exploit new technology;
4. To protect overall market share;
5. To maintain the enterprise business reputation as a product/service innovator.

Lateral diversification occurs when an enterprise business chooses to introduce new products to new customers. This approach to closing the performance gap is distinct in the sense that when an enterprise business diversifies, it essentially moves out of its current lines of products or services and markets into new areas different from its core businesses. It is important to note that lateral diversification may be into related and unrelated areas. Related diversification may be in the form of backward, forward, and horizontal integration. Backward integration takes place when the enterprise business extends its activities towards its inputs such as suppliers of raw materials etc. in the same business. Forward integration differs from backward integration, in that the enterprise business extends its activities towards its outputs such as distribution etc. in the same business. Horizontal integration takes place when an enterprise business moves into businesses that are related to its existing activities.

When the right investments are made, value creation (and the avoidance of value destruction) occurs through the processes of innovation, improved market penetration, product/services diversification, market diversification, or lateral diversification.

11.3 Value Creation Through Sustaining and Disruptive Innovation

Successful enterprise businesses, no matter what the source of their capabilities, are pretty good at responding to evolutionary changes in their markets – what in "The Innovator's Dilemma" (Christensen, 1997, 2003), Clayton Christensen referred to as sustaining innovation. Where they run into trouble is in handling or initiating revolutionary changes in their markets, or dealing with disruptive innovation.

Disruptive innovations create an entirely new market through the introduction of a new kind of product or service, one that is actually worse, initially, as judged by the performance metrics that mainstream customers' value. Early personal computers, for example, were a disruptive innovation relative to mainframes and minicomputers. Personal computers were not powerful enough to run the computing applications that existed at the time they were introduced. These innovations were disruptive in that they did not address the next-generation needs of leading customers in existing markets. They had other attributes, of course, that enabled new market applications to emerge – and the disruptive innovations improved so rapidly that they ultimately could address the needs of customers in the mainstream of the market as well.

Disruptive innovations occur so intermittently that no enterprise business has a routine process for handling them. Furthermore, because disruptive products nearly always promise lower profit margins per unit sold and are not attractive to the enterprise business' best customers', they are inconsistent with the established enterprise business values.

Sustaining technologies are innovations that make an enterprise business product or service perform better in ways that customers in the mainstream market already value. These are breakthrough innovations that sustained the best customers of these companies by providing something better than had previously been available.

Sustaining innovations are nearly always developed and introduced by established industry leaders. But those same enterprise businesses never introduce – or cope well with – disruptive innovations. Why? The resources-processes-values framework introduced earlier in the "Enterprise Culture & Values" section holds the answer. Industry leaders are organized to develop and introduce sustaining technologies. Month after month, year after year, they launch new and improved products to gain an edge over the competition. They do so by developing processes for evaluating the technological potential of sustaining innovations and for assessing their customers' needs for alternatives. Investment in sustaining technology also fits in with the values of leading companies in that they promise higher margins from better products sold to leading-edge customers.

Innovation in enterprise businesses is not only about product development. Innovation (sustaining or disruptive) is ultimately about finding ways to deliver new value to the marketplace from existing enterprise resources, whether this value is in the form of products, new work practices, improved processes, new management techniques or new business models. Innovation is an important partner to change. It is the wellspring of social and economic progress, and both a product and a facilitator of the free exchange of ideas that is the lifeblood of progress. It is reflected in new products, work practice, and work/production processes, advances in communications technology, etc.

As the systems model suggests, a successful enterprise businesses at the "Continuous Improvement" stage of maturity is likely to generate substantial surplus funds which are not required in order to maintain its leading position. How these funds get invested to continuously create value to its shareholders, customers, and employees, and captures value from its diverse resources and assets depends very much on the circumstances of the individual enterprise business.

In terms of goal orientation, the mature enterprise business, like many other systems, seeks to control its environment in order to ensure its own future well-being. The one thing the mature enterprise business, with all its heavy investments sunk in the status quo, seeks to avoid is change and instability. It will find it advantageous to invest heavily in preserving its current leader position within the current state of affairs.

Indeed, at the "Continuous Improvement" stage of maturity, it is difficult, however, for a successful enterprise business to make a fundamental change in what has established its leading position. This is especially the case with product and technology when a successful mature business is likely to have major investments sunk in the old product and technology. Getting into something fundamentally new may mean writing off huge capital assets which will weaken the balance sheet and in the short term wreak havoc with profitability. Also, there are non negligible psychological investments in the old product and technology used or developed by the enterprise business.

Furthermore, one of the fruits of maturity is the ability to pay top salaries and thus to attract highly qualified professionals. Many of these highly qualified professionals may have built their entire careers on the old product and technology and their very natural response to such a fundamental change is likely to be defensive and reactive. Nor is it at all certain that leadership in the new product and technology will necessarily follow being a leader in the old; giving up a leading position should certainly not be done lightly. Enterprise business executives and managers must consider the business risk exposure to the full range of macroeconomic and industry trends that will shape business performance.

While there are different types of risk that an enterprise business should consider (Duckert, 2010; Hampton, 2009; Hiles, 2004; Lam, 2003; Monahan, 2008), three overarching themes are especially important today:

1. *Macroeconomic risk* – It relates to the risk associated with uncertainties and turbulent in the macroeconomic environment. What would happen if a major growth market, say China, India or Brazil, enters a period of extended inflation or even stagnation? What would happen if the enterprise local currency weakens (or strengthens) significantly relative to the currencies of served markets or key sourcing regions?
2. *Capital-Market risk* – It relates to the risk associated with the current state of the capital markets. For example, many enterprise businesses are trading at high valuation multiples because, although their post downturn profits remain abnormally low, investors have already priced economic recovery into their

stock price. What would happen if those high multiples cannot be sustained? What would happen if the enterprise multiple relative to its peers starts to decline?

3. *Regulatory risk* – It relates to the risk associated with the growing role of regulatory governments. What is the possibility that new, more stringent, government regulations will constrain the enterprise ability to create value? Will new regulations or higher taxes seriously erode the enterprise ability to fund growth and innovation or continue its current level of payouts to investors?

Once the relevant uncertainties and turbulence in the macro-economy, the capital-market, and the regulatory environment have been identified, and once the potential risk has been quantified, successful enterprise businesses innovate through appropriate and relatively small incremental changes, as their businesses are able to focus, with the least inhibition and interruption, on the achievement of the enterprise intended strategy.

11.3.1 Creating Capabilities to Cope with Disruptive Innovation

Despite beliefs spawned by popular change-management and reengineering programs, processes – within the resources-processes-values framework introduced earlier in the "Enterprise Culture & Values" section – are not nearly as flexible or adaptable as resources are – and values are even less so. So whether addressing sustaining or disruptive innovations, when an enterprise business needs new processes and values – because it needs new capabilities – enterprise business executives must create a new organizational space where those capabilities can be developed. There are three possible ways to do that (Christensen & Overdorf, 2010). Enterprise business executives can:

1. Create new organizational structures within corporate boundaries in which new processes can be developed;
2. Spin out an independent organization from the existing organization and develop within it the new processes and values required to solve the new problem;
3. Acquire a different organization whose processes and values closely match the requirements of the new task.

11.3.2 Creating New Capabilities Internally

When an enterprise business' capabilities reside in its processes, and when new challenges require new processes – that is, when they require different people or groups in the enterprise business to interact differently and at a different pace than they habitually have done – managers need to pull the relevant people out of the existing enterprise business and draw a new boundary around a new group. Often, organizational boundaries were first drawn to facilitate the operation of existing processes, and they impede the creation of new processes. New team boundaries facilitate new patterns of working together that ultimately can coalesce as new processes. These structures are often referred to as "heavyweight teams."

These teams are entirely dedicated to the new challenge introduced by the disruptive change, team members are physically located together, and each member is charged with assuming personal responsibility for the success of the entire project. At Opel, for example, the boundaries of the groups within its product development organization historically had been defined by components – power train, electrical systems, and so on. But to accelerate auto development, Opel needed to focus not on components but on automobile platforms – small car for example – so it created heavyweight teams. Although these organizational units are not as good at focusing on component design, they facilitated the definition of new processes that were much faster and more efficient in integrating various subsystems into new car designs.

11.3.3 Creating Capabilities Through a Spinout Organization

When the mainstream enterprise business' values would render it incapable of allocating resources to an innovation project, the enterprise business should spin it out as a new venture. Large enterprise businesses cannot be expected to allocate the critical financial and human resources needed to build a strong position in small, emerging markets. And it is very difficult for an enterprise business whose cost structure is tailored to compete in high-end markets to be profitable in low-end markets as well. Spinouts are very much in vogue among managers in old-line enterprise businesses struggling with the question of how to address the Internet. But that is not always appropriate. When a disruptive innovation requires a different cost structure in order to be profitable and competitive, or when the current size of the opportunity is insignificant relative to the growth needs of the mainstream enterprise business, then – and only then – is a spinout organization required.

How separate does such an effort need to be? Clayton M. Christensen shows that a new physical location is not always necessary (Christensen & Overdorf, 2010). The primary requirement is that the project not be forced to compete for resources with projects in the mainstream enterprise business. Projects that are inconsistent with an enterprise business' mainstream values will naturally be accorded lowest priority. Whether the independent organization is physically separate is less important than its independence from the normal decision-making criteria in the resource allocation process.

Enterprise business executives often think that developing a new operation necessarily means abandoning the old one, and they are despise to do that since it works perfectly well for what it was designed to do. But when disruptive innovation appears on the horizon, enterprise business executives need to assemble the capabilities to confront that change before it affects the mainstream business. They actually need to run two businesses in tandem – one whose processes are tuned to the existing business model and another that is geared toward the new model.

11.4 Value Creation Through Diversification

In terms of goal orientation, the mature enterprise business, like many other systems, seeks to control its environment in order to ensure its own future well-being. The one thing the mature enterprise business, with all its heavy investments sunk in the status quo, seeks to avoid is change and instability. It will find it advantageous to invest heavily in preserving its current leader position within the current state of affairs.

First of all, it will seek to control its own industry, if possible through the achievement of a monopolistic position. In this way, it can hope to control the level of business to set limits on the competitive activity which would be profitable for any other business to embark on.

For similar reasons it may well seek to achieve control over its inputs whether sources of raw material or possibly core technologies, if they are in any way insecure. It may seek to achieve this through the exercise of its purchasing muscle in tying up long-term arrangements or by acquiring key suppliers.

In today's economy, however, successful enterprise businesses are constrained by anti-trust legislations which limit their ability to dominate their markets, prevent competition-reducing acquisitions and sometimes result in enterprise businesses being broken up if they became too dominant by organic growth. Growing enterprise businesses are therefore forced to diversify into new markets and new activities.

Under the diversification practice, the enterprise seeks to increase profitability through greater sales volume obtained from new products or services and new markets, as indicated in the previous section.

Diversification not only adds complexity and confuses strategic direction, but it also creates an alternative focus for future development and an ongoing demand for investment which may in the end result in the starvation of the once successful mature business. Thus diversification leads naturally to the mature business which has lost its direction and failed to keep up investment in its key businesses.

Conclusion 12

This book is intended to provide guidance to enterprises management and to professionals engaged in the "Continuous Improvement" initiative or program implementation and enable them to structure and manage its strategic implementation successfully. As you have realized, this is not a traditional book on Kaizen philosophy or practices that focus upon continuous improvement of processes in manufacturing, and engineering!

Throughout the chapters of this book, we have emphasized that successful enterprise businesses have to balance two needs:

1. The need to look backward in order to maintain the existing business and its current customers, and
2. The need to look forward in order to explore and achieve performance breakthroughs and to identify and attract new customers and new sources of value.

Achieving this balance requires specific and intensive actions along eight overarching determining factors, that matter the most, among dozens:

1. Leadership
2. Culture and Values
3. Strategic Planning and Management
4. Performance Measurement
5. Performance Management
6. Alignment and Commitment
7. Process Improvement and Management
8. Sustainability

There are literally hundreds of other books that emphasize the technical aspects of each of these overarching determining factors. The focus in these chapters has been on their strategic, social and organizational aspects that are so crucial to moving an enterprise business from its current maturity stage to a higher and ultimately to the "Continuous Improvement" maturity stage.

As exciting as these eight overarching determining factors are, they all have been strongly resisted, and they and others like them will continue to be resisted by

A. Van Aartsengel and S. Kurtoglu, *A Guide to Continuous Improvement Transformation*, Management for Professionals, DOI 10.1007/978-3-642-35904-0_12,

enterprise businesses that lack critical enablers. The type of transformation to a higher maturity stage that we advocate in this book will only come into being through multiple changes in the Context, Focus, Integration, and Interactivity of each of these overarching determining factors. By this, we extend the results of Dean Spitzer on "Transforming Performance Measurement" (Spitzer, 2007).

Without these enablers, the development and implementation of a "Continuous Improvement" initiative in your business does not stand much of a chance to succeed. As you have seen throughout the previous chapters, it is the strategic, social and organizational aspects of these eight overarching determining factors that impede real and deep change, and it is in transforming these aspects that the promise of "Continuous Improvement" maturity stage can be realized.

Throughout these chapters, we have outlined some practical steps that you, as enterprise business executive, manager or leader, can take right now to begin moving your business from its current maturity stage to a higher one. Do not be concerned if you cannot do all of these things, illustrated in Fig. 12.1, or even very many of them. The most successful transformations, especially those that touch virtually every aspect of an enterprise business performance like "Continuous Improvement" maturity stage does, are more likely to be evolutionary than revolutionary.

The inner wheel in Fig. 12.1 describes a basic improvement process. The outer wheel is intended to focus on how effectively your enterprise business goes about the business of improvement. How well do you do it? Do you do a better job of improving than your competition? The outer wheel focuses on the eight overarching determining factors of strategic management.

The previous chapters, as you have also realized, are not about the technical aspects of each of the eight overarching determining factors; they are about creating an optimal environment (context) for their effective use. Many enterprise businesses are engaged in implementing improvement initiatives with direct focus upon revenues, expenses and profits, but few are able to establish the right environment for the effective deployment of each of the eight overarching determining factors.

If your improvement activities are not ultimately providing more funds (i.e., bottom line cash flow and profits statements) for fulfilling your vision, then how effective can they be? As an enterprise executive, you must avoid the temptation to focus on cost reductions only, which is where the vast majority of your competitors focus. Instead, review the eight overarching determining factor of strategic management discussed throughout this chapter, and see how improving one or more of them might make a meaningful improvement in your business performance improvement recipe.

Most enterprise businesses have very competent people crunching to improve the business and have made massive investments to educate and train their employees on the technical aspects of each of the eight overarching determining factors. But almost none have done anything to improve the social and organizational context to make the improvement of these determining factors enjoyable and productive for all the people in the enterprise business.

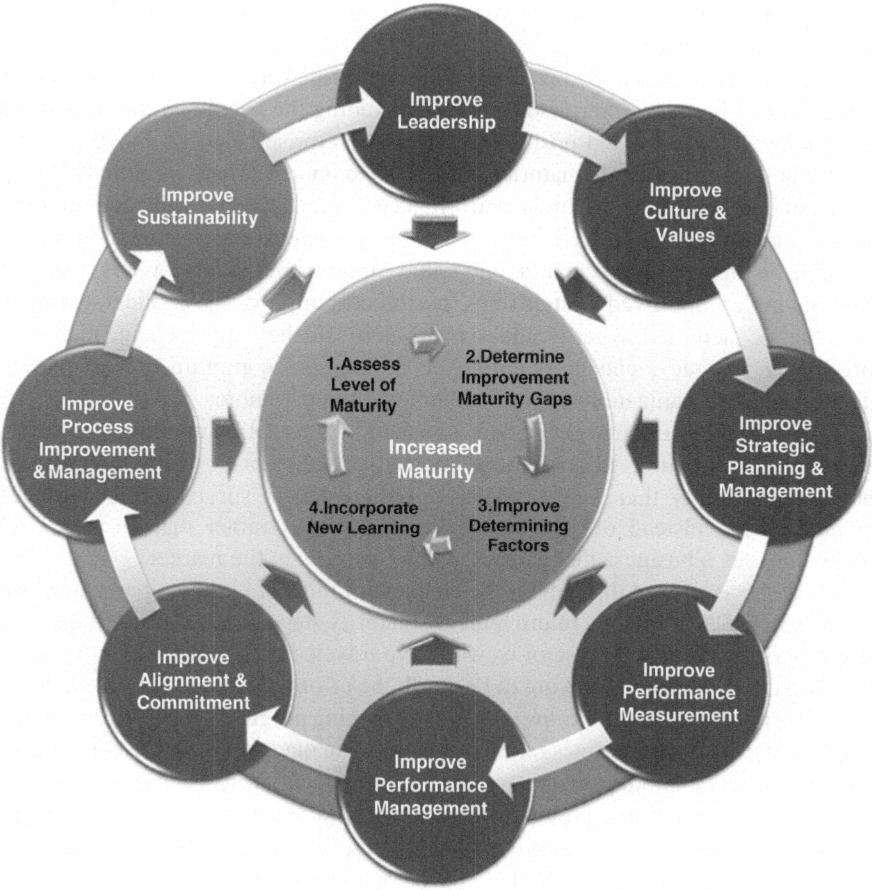

Fig. 12.1 The improvement maturity pathway to enterprise business excellence

Enterprise business executives and managers are expected to address their shareholders wealth, earnings growth, and total return on assets, but the most successful enterprise businesses understand that those measures should not be the primary targets of their intended strategies. Achieving attractive financial performance is the reward for having aimed at (and hit) the real target; i.e., maximizing the value created for the primary constituents of the enterprise business.

When an enterprise business thinks of itself as a financial engine whose purpose is to generate attractive financial returns with direct focus upon revenues, expenses and profits, the enterprise business is least likely to maximize those returns in the long run. Often, finance people end up shuffling a portfolio of assets in a self-destructive quest for "growth businesses" with no real understanding of the value-creation dynamics of enterprise businesses they are responsible for. Or, as with the banking or automotive service chains, attempts to profit without delivering superior value tends to place different corporate, their associated divisions and departments in competition. As a result, it ends in lost business, long-term customer alienation, and corporate disgrace.

When the right actions are taken and done right – the way that is being advocated in this book – both the enterprise business (including the people within it) and the customers will be impacted positively. For the enterprise business, moving to the "Continuous Improvement" maturity stage leads to improvements in virtually every aspect of the enterprise business performance from Accident reduction to Zero defects, including: improved intended strategy execution, better investment decisions, increased value creation and value capture from diverse assets (tangible and intangible), improved relationships (customers, employees, suppliers, partners, and others), increased synergy and synchronicity of the supply chain, increased forecasting accuracy, enhanced employee motivation, commitment and performance, greater organizational learning, and much, much more.

By encouraging internal cooperation instead of internal competition, an enterprise at the "Continuous Improvement" maturity stage releases within the business the creative energy that would otherwise be devoted to survival. It encourages communication instead of secrecy, which builds trust rather than distrust, and thereby creates a business that can respond to changes in the market.

As you have learned throughout the chapters of this book, moving an enterprise business to the "Continuous Improvement" maturity stage is a journey. It requires a new way of thinking and those of us who have traveled down this road universally proclaim the new way of thinking to be worth the effort, and the results do show up on the bottom line. You will know that you are making progress on the journey when people in the enterprise business start asking the right questions, and are engaging in dialogue rather than independent technical improvement projects or workshops.

We have indicated in a previous chapter that improvement in each of the eight overarching determining factors is about "change in form, quality, or state, over time." We have also indicated that the resources that support the work being completed to improve these determining factors or their constituent elements can be optimized methodically. Our next book entitled "Handbook on Continuous Improvement Transformation: The Lean Six Sigma Framework and Systematic Methodology for Implementation," will guide you through the technical methodic equally vital for implementing and sustaining a "Continuous Improvement" initiative or program in your enterprise business.

When speaking to, and consulting with, organization clients on: how to create awareness on "Continuous Improvement" transformation using the strategic management maturity model the way that we advocate in this book; how to have managers and staff develop the capability for ongoing, incremental evolution to maturity and improvement; we have found that they often do not know where to start or what they should be aiming to achieve. Enterprise businesses realize that they have a long way to go to achieve a "Continuous Improvement" transformation but have no overall structured approach of how to get there and what steps they need to take. This journey to the "Continuous Improvement" state of maturity is a large and complicated set of tasks for an enterprise business, so we have broken the journey down into eight overarching dimensions of strategic management that matter the most. We have then provided, for each dimension, a visionary state of maturity and then provide a practical roadmap of how to get there.

Based on the material that we have provided throughout this book, the challenge is not to turn the heads of enterprise business leaders – from CEOs to supervisors – toward implementing the eight overarching determining factors of strategic management selected and developed, but to manage repeatedly and consistently those determining factors strategically, synergistically, and with appropriate alignment and synchronicity to achieve systematic maturation and improvement across the enterprise business. As we have indicated already, the maturation process is about movement to a higher stage of being, whereas improvement in each overarching determining factor of strategic management is about "change in form, quality, or state, over time."

Within enterprise businesses at a "Continuous Improvement" maturity stage like Toyota's and GE's family of companies, observation of employees behavior shows that people feel good about focusing on developing daily behavior patterns by sensing and understanding any improvement opportunity and reacting to it in a way that moves the enterprise business forward. Instead of an activity to be avoided or feared, "Continuous Improvement" transformation thinking pattern becomes an activity that employees actually enjoy, and opportunity for business activities improvement become an activity that employees look forward to performing. Within these families of companies, there is no "finish line" mentality. The objective is not to win, but to develop the capability of the enterprise business to keep improving, adapting, and satisfying dynamic customer requirements. This capability for ongoing, incremental evolution and improvement represents the best assurance of durable competitive advantage and enterprise business survival.

12.1 Improving and Managing

Within an enterprise business at the "Continuous Improvement" stage of maturity, improving and managing are one and the same. The framework and systematic methodology provided in our next book is to a considerable degree how such an enterprise manages its activities, processes and people from day to day. In comparison, enterprise businesses at lower maturity stage tend to see managing as a unique and separate activity. Improvement is something extra, added on to managing.

An interesting point here is that many of managers would probably be afraid to focus so heavily on this second philosophy (i.e. improvement) at the expense of the first philosophy (i.e. make production). They would feel that they are letting go of something they currently try very hard to control, because they are accustomed to focusing on outcomes, not on drivers and process details. In such approach managers concentrate on outcome targets and consequences. And consequently, the enterprise business will think of itself as a financial engine whose purpose is to generate attractive financial returns with direct focus upon

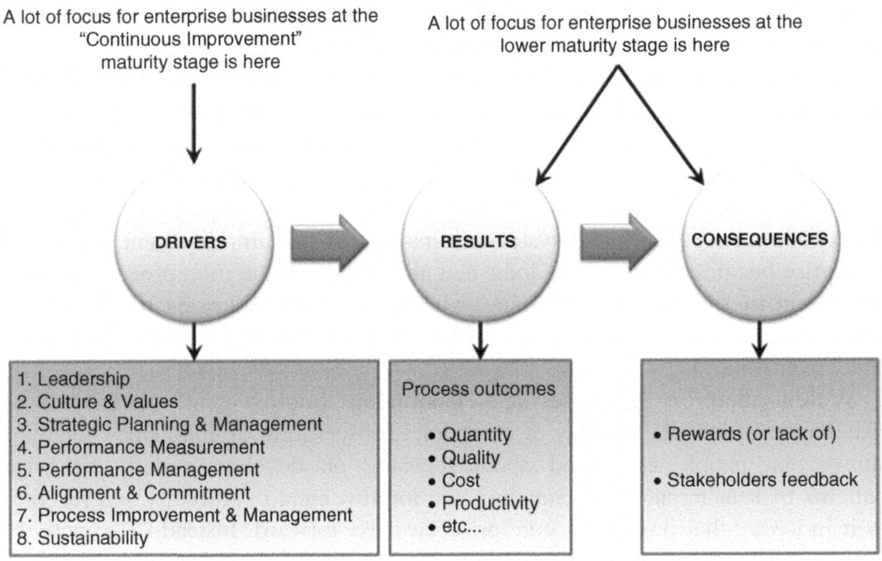

Fig. 12.2 Focusing on means in order to achieve desired results

revenues, expenses and profits. As such, it is least likely to maximize those returns in the long run.

In contrast, as depicted in Fig. 12.2, an enterprise business at the "Continuous Improvement" stage of maturity puts considerable emphasis on how to tackle the drivers and details of a process, which is what generates the outcomes.

Outcome targets, such as the desired production quantity and quality, are of course necessary. But if you focus on continuously improving the process – systematically, through the framework and systematic methodology provided in this book, rather than just random improvement – then the desired outcomes will come. Making the desired production or service quantity and quality, for example, will happen automatically when you focus on the drivers and details of a process through correct application of the systematic methodology provided.

The eight overarching determining factors of strategic management listed in the figure above are drivers necessary to achieve a state of "Continuous Improvement" as developed in this book. Each of the drivers can be considered at both operational and departmental level within an enterprise business if each department is considered as a unit which acquired and provide services and products to the other department within an enterprise business. The relative importance of each driver will vary from situation to situation and from time to time depending upon market pressures and other changing circumstances but it is unlikely that the total number will be fewer than the eight overarching determining factors of strategic management shown in Fig. 12.2 above. It is best to play safe and consider all of these eight determining factors all of the time. Conceivably, there may be more in some situations but in many years of experience with a range of businesses across a spectrum of industries we have not yet found such a case.

12.2 Final Admonition

As we consider the enterprise maturity, from the perspective of systems theory, it is important to remember that "Continuous Improvement" maturity is not about strength in any one aspect of the enterprise, but about the health of the enterprise as a "total system." Some enterprise businesses excel at one, two or three of the eight determining factors of strategic management elaborated. However, for any enterprise to achieve superior results, it is essential that all eight determining factors work in tandem with each other.

If your enterprise business falls behind your competitors, it is generally not possible to catch up quickly or in a few leaps. If there was something you could do, or implement, to get caught up again quickly, then your competitors will be doing that too. If you want your enterprise business to thrive for a long time, then how it interacts continuously with conditions inside and outside the enterprise business through the eight overarching determining factors of strategic management is important. This capability for ongoing, incremental evolution to maturity and improvement represents the best assurance of durable competitive advantage and company survival, because:

1. *Small, incremental steps let you learn along the way, make adjustments, and discover the path to where you want to be*. Since you cannot see very far ahead, you cannot rely on up front planning alone. Improvement, adaptation, and even innovation result to a great extent from the accumulation of small steps; each lesson learned helps you recognize the next step and adds to your knowledge and capability.
2. *Relying on technical innovation alone often provides only temporary competitive advantage*. Technological innovations are important and offer competitive advantage, but they come infrequently and can often be copied by competitors. In many cases you cannot expect to enjoy more than a brief technological advantage over competitors. Technological innovation is also arguably less the product of revolutionary breakthroughs by single individuals than the cumulative result of many incremental adaptations and improvements that have been pointed in a particular direction and conducted with special focus and energy.
3. *Cost and quality competitiveness tend to result from accumulation of many small steps over time*. Again, if one could simply implement some measures to achieve cost and quality competitiveness, then every enterprise business would do it. Cost and quality improvements are actually made in small steps and take considerable time to achieve and accumulate. The results of ongoing and small steps cost reduction and quality improvement are therefore difficult to copy, and thus offer a special competitive advantage. It is highly advantageous for an enterprise business in a competitive environment to combine efforts at innovation with unending ongoing improvement of cost and quality competitiveness, even in the case of mature products or services.
4. Relying on periodic improvements and innovations alone – only improving when you make a special effort or campaign – conceals a system that is static

and vulnerable. Here is an interesting point to consider about your own enterprise business: in many cases the normal operating condition of an enterprise business – its nature – is not improving. Many enterprise business executives and managers think of improvement as something that happens periodically, like a project or campaign: a special effort to improve or change is made when the need becomes urgent. But this is not how a "Continuous Improvement" state of maturity actually comes about. Relying on periodic improvement or change efforts should be seen for what it is: only an occasional add-on to a system that by its nature tends to stand still.

Enterprise businesses that understand and can use the material described in this book to manage their strategy, systems, and processes more effectively and more consistently will find that it provides a tremendous competitive advantage. Successful incumbents must tolerate initial failure and grasp the need for course correction. In effect, enterprise businesses have to focus on learning and adjusting as much as on executing.

We urge enterprise businesses with new business models for "Continuous Improvement" to be patient for realizing "Continuous Improvement" capability but impatient for realizing short term profit as an early validation that the model works. A short term profitable business is the best early indication of a viable model.

Real transformation takes time, and a renewal effort risks losing momentum if there are no short-term goals to meet and celebrate. Most people won't go on the long march unless they see compelling evidence within 1 or 2 years that the journey to "Continuous Improvement" is producing expected results. Without short-term profits, too many people give up or actively join the ranks of those people who have been making waves.

One to two years into a successful transformation effort, you find quality beginning to go up on certain indices or the decline in net income stopping. You find some successful new product or service introductions or an upward shift in market share. You find an impressive productivity improvement or a statistically higher customer-satisfaction rating. But whatever the case, the profit is unambiguous. The result is not just a judgment call that can be discounted by those opposing the transformation to "Continuous Improvement" stage of maturity.

Creating short-term profits is different from hoping for short-term profits. The latter is passive, the former active. In a successful transformation, enterprise business executives and managers actively look for ways to obtain clear performance improvements, establish goals in the yearly planning system, achieve the objectives, and reward the people involved with recognition and promotion. Enterprise business executives and managers often complain about being forced to produce short-term profits, but we have found that pressure can be a useful element in a transformation effort to "Continuous Improvement" stage of maturity. When it becomes clear to people that major transformation will take a long time, urgency levels can drop. Commitments to produce short-term profits help keep the urgency level up and force detailed analytical thinking that can clarify or revise visions.

The progressive realization of the enterprise full potential by moving from its current maturity stage towards a higher (ultimately "Continuous Improvement") maturity stage, requires a framework and a systematic methodology for studying the constituent elements or processes and systems associated with the eight overarching determining factors. It also requires a way of differentiating between the different types of variation present in those processes and systems. In addition to the way of thinking described throughout the chapters of this book, and which must be put to practice, there are techniques to be learned. In our next book, we will describe the framework and systematic methodology for improving processes used within projects and operations work.

Bibliography

Aaker, D. A. (1992). *Developing business strategies*. New York: Wiley.

Abraham, S. C. (2012). *Strategic planning: A practical guide for competitive success*. Bradford: Emerald Group Publishing.

Adair, J. (1997). *Decision making and problem solving*. London: CIPD Publishing.

Adair, J. E. (2009). *Effective decision making (Rev. Ed.): The essential guide to thinking for management success*. London: Pan Macmillan.

Akao, Y. (2004). *QFD: Quality Function Deployment – Integrating customer requirements into product design*. Cambridge, MA: Productivity Press.

Alvesson, M. (2002). *Understanding organizational culture*. London: Sage.

Ansoff, H. I. (1965). *Corporate strategy: An analytic approach to business policy for growth and expansion*. New York: McGraw-Hill.

Ansoff, H. I. (1976). *From strategic planning to strategic management*. New York: Wiley.

Ashkanasy, N. M., Wilderom, C. P., & Peterson, M. F. (2010). *The handbook of organizational culture and climate*. Thousand Oaks, CA: Sage.

Barksdale, S., & Lund, T. (2006). *10 steps to successful strategic planning*. Washington, DC: American Society for Training and Development.

Bell, D. E., Raiffa, H., & Tversky, A. (1988). *Decision making: Descriptive, normative, and prescriptive interactions*. Cambridge, UK: Cambridge University Press.

Bergiel, E. B. (2006). *Shared mental models and team performance: Clarifying the group process mediator of cohesion*. Ph.D. thesis, Mississippi State University. Department of Management and Information Systems.

Bhardwaj, J. (2010). *Application of quality function deployment in product development*. Germany: LAP Lambert Academic Publishing AG & Co KG.

Bossert, J. L. (1991). *Quality function deployment: A practitioner's approach*. Milwaukee, WI: ASQC Quality Press.

Bossidy, L., Charan, R., & Burck, C. (2002). *Execution: The discipline of getting things done*. New York: Crown Business.

Bradford, R. W., Duncan, J. P., & Tarcy, B. (2000). *Simplified strategic planning: A no-nonsense guide for busy people who want results fast!* Worcester, MA: Chandler House Press.

Brannick, M. T., Salas, E., & Prince, C. W. (1997). *Team performance assessment and measurement: Theory, methods, and applications*. New York: Routledge.

Brenton, A. L., & Driskill, G. W. (2010). *Organizational culture in action: A cultural analysis workbook*. Thousand Oaks, CA: Sage.

Brest, P., & Krieger, L. H. (2010). *Problem solving, decision making, and professional judgment*. Oxford, UK: Oxford University Press.

Bridge, J., & Dodds, J. C. (1975). *Managerial decision making*. New York: Press Taylor & Francis Group.

BSI, B. S. (2010). *The strategic management maturity model*. Retrieved from balancedscorecard.org: http://www.balancedscorecard.org/Portals/0/PDF/BSCIStrategicManagementMaturityModel.pdf

A. Van Aartsengel and S. Kurtoglu, *A Guide to Continuous Improvement Transformation*, Management for Professionals, DOI 10.1007/978-3-642-35904-0,

Burke, L. A., & Miller, M. K. (1999). Taking the mystery out of intuitive decision making. *Academy of Management Executive, 13*(4), 91–98.

Cameron, K. S., & Quinn, R. E. (2011). *Diagnosing and changing organizational culture: Based on the competing values framework*. New York: Wiley.

Cannon-Bowers, J. A., & Salas, E. (2001). Team effectiveness and competencies. In W. Karwowski (Ed.), *International encyclopedia of ergonomics and human factors* (pp. 1383–1384). London/New York: Taylor & Francis.

Chambers, L., & Taylor, M. A. (1999). *Strategic planning: Processes, tools and outcomes*. Aldershot, UK: Ashgate Publishing.

Chen, M. (2006). Understanding the benefits and detriments of conflict on team creative process. *Creativity and Innovation, 15*(1), 105–116.

Christensen, C. M. (1997). *The innovator's dilemma when new technologies cause great firms to fail*. Boston: Harvard Business School Press.

Christensen, C. M. (2003). *The innovator's dilemma: The revolutionary book that will change the way you do business (Collins Business Essentials)*. Scarborough, ON: Harper Paperbacks.

Christensen, C. M., Anthony, S. D., & Roth, E. A. (2004). *Seeing what's next: Using the theories of innovation to predict industry change*. Boston: Harvard Business Press.

Christensen, C. M., & Overdorf, M. (2010). *Meeting the challenge of disruptive change*. Boston: Harvard Business School Publishing Corporation.

Cohen, L. (1995). *Quality function deployment: How to make QFD work for you*. Englewood Cliffs, NJ: Prentice Hall.

Craik, K. J. (1967). *The nature of explanation*. Cambridge, UK: Cambridge University Press.

Day, R. G. (1993). *Quality function deployment: Linking a company with its customer*. Milwaukee, WI: ASQC Quality Press.

De Dreu, C. K., & Weingart, L. R. (2003a). Task versus relationship conflict: Team performance, and team member satisfaction. *Journal of Applied Psychology, 88*(4), 741–749.

De Dreu, C., & Weingart, L. (2003b). Task versus relationship conflict and team effectiveness: A meta-analysis. *Journal of Applied Psychology, 88*, 741–749.

Deal, T. E., & Kennedy, A. A. (1982). *Corporate cultures: The rites and rituals of corporate life*. Reading, MA: Addison-Wesley.

Deal, T. E., & Kennedy, A. A. (2000). *Corporate cultures: The rites and rituals of corporate life*. London: Perseus Books.

Drucker, P. F. (1964). *Managing for results*. London: Heinemann Professional Publishing.

Duckert, G. H. (2010). *Practical enterprise risk management: A business process approach*. Hoboken, NJ: Wiley.

Duffy, G. L., Moran, J. W., & Riley, W. (2010). *Quality function deployment and lean-six sigma applications in public health*. Milwaukee, WI: ASQ Quality Press.

Espy, S. N. (1986). *Handbook of strategic planning for nonprofit organizations*. New York: Greenwood Publishing Group.

Fairholm, M. R., & Fairholm, G. W. (2008). *Understanding leadership perspectives: Theoretical and practical approaches*. New York: Springer.

Ficalora, J. P., & Cohen, L. (2010). *Quality function deployment and six sigma: A QFD handbook*. Upper Saddle River, NJ: Prentice Hall.

Fogg, C. D. (1994). *Team-based strategic planning: A complete guide to structuring, facilitating, and implementing the process*. New York: AMACOM: A Division of American Management Association.

Gettman, D. J. (2001). *An investigation of the shared mental model construct in the context of a complex team task*. Unpublished doctoral dissertation, Texas A & M University.

Goleman, D. (2007). *Social intelligence*. New York: Bantam Books.

Goodstein, L. D., Nolan, T. M., & Pfeiffer, J. W. (1993). *Applied strategic planning: A comprehensive guide*. New York: McGraw-Hill Professional.

Gregory, K. (1983). Native-view paradigms: Multiple cultures and culture conflicts in organisations. *Administrative Science Quarterly, 28*, 359–376.

Grünig, R., Kühn, R., & Clark, A. (2010). *Process-based strategic planning*. Heidelberg: Springer.

Hackman, J. R. (1976). Group influences on individuals. In M. D. Dunnette (Ed.), *Handbook of industrial and organizational psychology*. Chicago: Rand McNally College Publishing Co.

Hampton, J. J. (2009). *Fundamentals of enterprise risk management: How top companies assess risk, manage exposures, and seize opportunities*. New York: AMACOM: A Division of American Management Association.

Harvard Business School. (2006). *Decision making: 5 steps to better results*. Boston: Harvard Business Press.

Harvard Business School. (2008). *Making decisions: Expert solutions to everyday challenges*. Boston: Harvard Business Press.

Hiles, A. (2004). *Enterprise risk assessment and business impact analysis: Best practices*. USA, 4 Arapaho Road Brookfield, CT 06804–3104.

Hughes, M., & Terrell, J. (2008). *The emotionally intelligent team*. San Francisco: Jossey-Bass.

Jeston, J., & Nelis, J. (2008a). *Business process management: Practical guidelines to successful implementations*. Burlington, MA: Butterworth-Heinemann.

Jeston, J., & Nelis, J. (2008b). *Management by process: A roadmap to sustainable business process management*. Amsterdam/Boston: Elsevier/Butterworth-Heinemann.

Jönsson, S. (1996). *Decoupling hierarchy and accountability: An examination of trust and reputation*. London: Thomson Business Press.

Kaplan, R. S., & Norton, D. P. (1996). *The balanced scorecard: Translating strategy into action*. Boston: Harvard Business School Press.

Katzenbach, J. R., & Smith, D. K. (1993). *The wisdom of teams: Creating the high-performance organization*. Boston: Harvard Business School.

Kaufman, R. A., Oakley-Browne, H., & Watkins, R. (2003). *Strategic planning for success: Aligning people, performance, and payoffs*. San Francisco: Wiley.

Koehler, W. (1938). *Closed and open systems*. Harmondsworth, UK: Penguin Books, Penguin Modern Management Readings.

Koehler, D. J., & Harvey, N. (2004). *Blackwell handbook of judgment and decision making*. Malden, MA: Wiley.

Kotter, J. P., & Heskett, J. L. (1992). *Corporate culture and performance*. New York: Simon and Schuster.

Lam, J. (2003). *Enterprise risk management: From incentives to controls*. Hoboken, NJ: Wiley.

Leibner, J., Mader, G., & Weiss, A. (2009). *The power of strategic commitment: Achieving extraordinary results through total alignment and engagement*. New York: AMACOM: A Division of American Management Association.

Letsky, M. P. (2008). *Macrocognition in teams: Theories and methodologies*. Aldershot, UK: Ashgate Publishing Ltd.

Lim, C. P. (2010). *Handbook on decision making: Vol 1: Techniques and applications*. Berlin/Heidelberg: Springer.

Lu, J., Jain, L. C., & Zhang, G. (2012). *Handbook on decision making: Vol 2: Risk management in decision making*. Berlin/New York: Springer.

MacMillan, P. (2001). *The performance factor: Unlocking the secrets of teamwork*. Nashville, TN: B&H Publishing Group.

Madu, C. N. (2000). *House of quality (QFD) in a minute: Quality function deployment*. Fairfield, CT: Chi Publishing.

Mann, D. W. (2005). *Creating a lean culture: Tools to sustain lean conversions*. New York: Productivity Press.

March, J. G. (1994). *Primer on decision making: How decisions happen*. New York: Simon and Schuster.

Martin, J. (2002). *Organizational culture: Mapping the terrain*. Thousand Oaks, CA: Sage.

McGrath, J. E. (1984). *Groups: Interaction and performance*. Englewood Cliffs, NJ: Prentice Hall.

McGrew, A. G., & Wilson, M. J. (1982). *Decision making: Approaches and analysis*. Manchester, UK: Manchester University Press ND.

Mintzberg, H. (1978). Patterns in strategy formation. *Management Science, 24*(9), 934–948.

Mintzberg, H. (1994). *The rise and fall of strategic planning: Reconceiving roles for planning, plans, planners*. New York: Simon and Schuster.

Misztal, B. A. (1996). *Trust in modern societies: The search for the bases of social order*. Cambridge, UK: Polity Press.

Mizuno, S., & Akao, Y. (1994). *QFD: The customer-driven approach to quality planning and development*. Tokyo: Asian Productivity Organization, Quality Resources.

Monahan, G. (2008). *Enterprise risk management: A methodology for achieving strategic objectives*. Hoboken, NJ: Wiley.

Moody, P. E. (1983). *Decision making: Proven methods for better decisions*. New York: McGraw-Hill.

Nutt, P. C., & Wilson, D. C. (2010). *Handbook of decision making*. West Sussex, UK: Wiley.

Olsen, E., Plaschke, F., & Stelter, D. (2008). *The 2008 value creators' report: Missing link focusing corporate strategy on value creation*. Amsterdam: Boston Consulting Group.

Olsen, E., Plaschke, F., & Stelter, D. (2009). *The 2009 value creators report – Searching for sustainability: Value creation in an era of diminished expectations*. Amsterdam: Boston Consulting Group.

Olsen, E., Plaschke, F., & Stelter, D. (2010). *The 2010 value creators report – Threading the needle: Value creation in a low-growth economy*. Amsterdam: Boston Consulting Group.

Olsen, E., Plaschke, F., Stelter, D., & Farag, H. (2011). *The 2011 value creators report – risky business: Value creation in a volatile economy*. Amsterdam: Boston Consulting Group.

Parker, G. (1994). *Crossfunctional teams*. San Francisco: Jossey-Bass.

Parker, M. (1999). *Organizational culture and identity: Unity and division at work*. London: Sage.

Peters, T. J., & Waterman, R. H. (2004). *In search of excellence: Lessons from America's best-run companies*. New York: HarperBusiness Essentials.

Pheysey, D. C. (1993). *Organizational cultures: Types and transformations*. New York: Routledge.

Rea, P. J., & Kerzner, H. (1997). *Strategic planning: A practical guide*. New York: Wiley.

ReVelle, J. B., Moran, J. W., & Cox, C. A. (1998). *The QFD handbook*. New York: Wiley.

Salas, E., & Klein, G. (2001). *Linking expertise and naturalistic decision making*. Mahwah, NJ: Lawrence Erlbaum.

Schein, E. H. (2009). *The corporate culture survival guide*. New York: Wiley.

Schein, E. H. (2010). *Organizational culture and leadership*. New York: Wiley.

Senge, P. M. (1994). *The fifth discipline: The art & practice of the learning organization*. New York: Doubleday Business.

Simerson, B. K. (2011). *Strategic planning: A practical guide to strategy formulation and execution*. Santa Barbara, CA: Praeger Publishers.

Smith, G. (2001). Group development: A review of the literature and a commentary on future research directions. *Group Facilitation, 3*, 14–45.

Spitzer, D. R. (2007). *Transforming performance measurement: Rethinking the way we measure and drive organizational success*. New York: AMACOM: A Division of American Management Association.

Stout, R. J., Cannon-Bowers, J. A., & Salas, E. (1996). The role of shared mental models in developing team situational awareness: Implications for training. *Training Research Journal, 2*, 85–116.

Stout, R. J., Cannon-Bowets, J. A., Salas, E., & Milanovich, D. M. (1999). Planning, shared mental models, and coordinated performance: An empirical link is established. *The Journal of the Human Factors and Ergonomics Society, 41*(1), 61–71.

Straw, J., & Cerier, A. B. (2002). *The 4-dimensional manager: Disc strategies for managing different people in the best ways*. San Francisco: Berrett-Koehler Publishers.

Sullivan, L. (1986). Quality function deployment. *Quality Progress, 19*(6), 39–50.

Teale, M. (2003). *Management decision making: Towards an integrated approach*. Harlow, UK: Pearson Education.

Terninko, J. (1997). *Step-by-step QFD: Customer-driven product design*. Nottingham, NH: CRC Press.

Tubbs, S. (1995). *A systems approach to small group interaction*. New York: McGraw-Hill.

Tuckman, B. W. (1965). Developmental sequence in small groups. *Psychological Bulletin, 63*(6), 384–399.

Tuckman, B. W., & Jensen, M. A. (1977). Stages of small-group development revisited. *Group and Organizational Studies, 2*, 419–427.

Weick, K., & Quinn, R. (1999). Organizational change and development. *Annual Review of Psychology, 50*, 361–386.

Welins, R., Byham, W., & Dixon, G. (1994). *Inside teams*. San Francisco: Jossey-Bass.

Wheelan, S. A. (1990). *Facilitating training groups: A guide to leadership and verbal intervention skills*. New York: Praeger.

Wheelan, S. A. (1994a). *Group processes: A developmental perspective*. Boston: Allyn & Bacon.

Wheelan, S. A. (1994b). *The group development questionnaire: A manual for professionals*. Provincetown, MA: GDQ Associates.

Wheelan, S., Davidson, B., & Tilin, F. (2003). Group development across time: Reality or illusion? *Small Group Research, 34*(2), 223–245.

Witte, K. D., & Muijen, J. V. (2000). *Organizational culture*. Milton Park, Abingdon, Oxford, OX14 4RN, UK: Psychology Press.

Wittmann, R., & Reuter, M. P. (2008). *Strategic planning: How to deliver maximum value through effective business strategy*. Philadelphia: Kogan Page.

Wootton, S., & Horne, T. (1997). *Strategic planning: The nine step programme: Putting theory into practice: A step-by-step approach*. London: Kogan Page.

Xie, M., Goh, T. N., & Tan, K. C. (2003). *Advanced QFD applications*. Milwaukee, WI: ASQ Quality Press.

Index

A. Van Aartsengel and S. Kurtoglu, *A Guide to Continuous Improvement Transformation*, Management for Professionals, DOI 10.1007/978-3-642-35904-0,

The manufacturer's authorised representative in the EU is Springer Nature Customer Service Centre GmbH, Europaplatz 3, 69115 Heidelberg, Germany. If you have any concerns regarding our products, please contact ProductSafety@springernature.com

Printed and bound by CPI Group (UK) Ltd, Croydon, CR0 4YY
25/09/2026
02222869-0004